Some Highlights

1. We show that tachyons exist within Black Holes.

2. We extend Special Relativity to include left-handed (Superluminal) transformations to reference frames moving at relative velocities greater than the speed of light. This set of transformations forms a group that includes the Lorentz group as an invariant subgroup. The speed of light does not change under the transformations of this group. We find new features in frames moving at a relative speed greater than the speed of light such as length *dilation*, time *contraction*, and tachyons decaying into more massive tachyons – "reverse fission."

3. Using the (Superluminal) transformations of this enlarged group we are able to define tachyons of half-integer or integer spin. We show these tachyons are local and satisfy canonical commutation relations in light-front coordinates (the "infinite momentum" frame). Thus a standard quantization procedure is possible and a conventional light-front perturbation theory can be constructed.

4. Free spin ½ tachyons violate parity and CPT but do not violate C or T invariance.

5. The requirement of Left-handed Extended Lorentz group covariance implies an extended Dirac equation for spin ½ particles with doublets of spin ½ particles: a Dirac particle and a tachyon. We identify neutrinos with tachyon members of lepton doublets, and "d-type" quarks with tachyon members of quark doublets.

6. Further considerations lead to most features of the Standard Model for one generation of leptons and quarks. Thus the general form of the Standard Model, including a rationale for the form of parity violation, $SU(2)\otimes U(1)$, and left-handed doublets and right-handed singlets, is derived.

7. The theory requires quark confinement. The theory does not explain the existence of three generations or the mixing of generations. SU(3) is found to be the minimal symmetry group of the strong interaction if spin ½ baryon bound states are to exist.

Physics Beyond the Light Barrier:
The Source of Parity Violation, Tachyons, and
A Derivation of Standard Model Features

Some Other Books by Stephen Blaha

A Derivation of ElectroWeak Theory based on an Extension of Special Relativity; Black Hole Tachyons; & Tachyons of Any Spin (ISBN: 0974695866, Pingree-Hill Publishing, Auburn, NH, 2006)

Quantum Theory of the Third Kind: A New Type of Divergence-free Quantum Field Theory Supporting a Unified Standard Model of Elementary Particles and Quantum Gravity based on a New Method in the Calculus of Variations (ISBN: 0974695831, Pingree-Hill Publishing, Auburn, NH, 2005)

Quantum Big Bang Cosmology: Complex Space-time General Relativity, Quantum Coordinates™ Dodecahedral Universe, Inflation, and New Spin 0, ½, 1 & 2 Tachyons & Imagyons™ (ISBN: 0974695815, Pingree-Hill Publishing, Auburn, NH, 2004)

The Metatheory of Physics Theories, and the Theory of Everything as a Quantum Computer Language (ISBN: 097469584X, Pingree-Hill Publishing, Auburn, NH, 2005)

A Unified Quantitative Theory Of Civilizations and Societies: 9600 BC - 2100 AD (ISBN: 0974685858, Pingree-Hill Publishing, Auburn, NH, 2006)

The Life Cycle of Civilizations (ISBN: 0972079580, Pingree-Hill Publishing, Auburn, NH, 2002)

Cosmos and Consciousness Second Edition (ISBN: 0972079548, Pingree-Hill Publishing, Auburn, NH, 2002)

Available on bn.com, Amazon.com, and other web sites as well as at better bookstores (through Ingram Distributors).

Cover Credits

Cover from a painting by Stephen Blaha © 2007.

Physics Beyond the Light Barrier: The Source of Parity Violation, Tachyons, and A Derivation of Standard Model Features

Stephen Blaha, Ph.D.*

Pingree-Hill Publishing

* Email: sblaha000@yahoo.com

Copyright © 2006-7 by Stephen Blaha
All rights reserved.

This document and related course software are protected under copyright laws and international copyright conventions. No part of this book may be reproduced, stored in a retrieval system, or transmitted by any means in any form, electronic, mechanical, photocopying, recording, or otherwise, without the express prior written permission of Pingree-Hill Publishing. For additional information email JanusAssc@yahoo.com or write to:

Pingree-Hill Publishing
P. O. Box 368
Auburn, NH 03032

ISBN: 0-9746958-7-4

This document is provided "as is" without a warranty of any kind, either implied or expressed, including, but not limited to, implied warranties of fitness for a particular purpose, merchantability, or non-infringement. This document may contain typographic errors, technical inaccuracies, and may not describe recent developments.

This book is printed on acid free paper.

rev. 00/00/02

Preface to the Revised and Expanded Version

This is a revised and expanded version of *A Derivation of ElectroWeak Theory based on an Extension of Special Relativity; Black Hole Tachyons; & Tachyons of Any Spin*. This book changes the extended Lorentz group called the Luminal group in the previous work to a different group – actually a pair of groups – that include the Lorentz group as a subgroup. These extended Lorentz groups each extend the Lorentz group to include transformations to faster-than light reference frames but with a major difference. One group (the Left-handed Extended Lorentz group) leads to the experimentally observed Standard Model, which is dominated by the left-handed particle interactions. The other group (the Right-handed Extende Lorentz group) leads to a contrary (unphysical) Standard Model, which is dominated by right-handed particle interactions.

Thus we reduce the observed parity asymmetry in Nature to the group structure of extended Special Relativity. The handedness of the Standard Model arises from the structure of the extended Lorentz group (the Left-handed Extended Lorentz group) in the sector defining faster-than-light transformations.

Extended Lorentz transformations have very different properties from Lorentz transformations. Some of the new phenomena associated with them include time *contraction*, length *dilation*, and particles decaying into *heavier* particles. (See Appendix 1-A.)

The tachyon theory of the previous version is largely unchanged. The derivation of Standard Model features is also largely unchanged.

PREFACE

Before Einstein the Newtonian universe looked like a simpler place: absolute (flat) space, simple transformations between reference frames, time and space did not mix, time had an absolute character and there was no limit on the speed of objects traveling through space.

Einstein originated a seemingly simple postulate: the speed of light is constant in all inertial reference frames and is independent of the speed of the source of the light. He added the postulate that the laws of physics in any inertial reference frame have the same form and do not depend on any absolute speed of motion either directly or indirectly. Einstein then derived the relation between the coordinates and times of two coordinate systems in constant relative motion, which we now call a *Lorentz transformation*. A study of Lorentz transformations suggests that particles with mass cannot have speeds greater than or equal to the speed of light because the energy of a particle approaches infinity as a particle's speed approaches the speed of light from a speed below light-speed. Thus Einstein concluded that particles with mass could not go faster than the speed of light.

Light itself, which is composed of massless particles called photons, travels at the speed of light in all reference frames as do any other massless particles.

Various physicists have considered the possibility of massive particles that travel faster than the speed of light. These proposed particles, called tachyons, have a finite energy and momentum. However previous studies of tachyon quantization have suggested tachyons have inherent problems, and the study of faster than light physics has languished.

In this book we will show that tachyons indeed exist – in Black Holes. And we will show the problems of tachyon quantization disappear if tachyon quantization is done properly – in light-front coordinates.

We will examine the physics of the "other half" of the universe – the part of the universe that "travels faster than the speed of light" – and show that it has a direct impact on the features of the Standard Model.

We will generalize Lorentz transformations to include transformations to reference frames moving faster than the speed of light. We call the combined group of transformations the Left-handed Lorentz Group. We will show that this group contains the Lorentz group as an invariant subgroup. We will then show how tachyons naturally appear in this extended formulation. We show how to define quantum field theories for tachyons of any integer or half-integer spin. The conditions of Wigner's theorem, which states that the only finite-dimensional representation of the Lorentz group for imaginary mass is the scalar representation, is not relevant for the Left-handed Extended Lorentz group. Thus spin ½ tachyons as well as higher spin tachyons are allowed.

We then derive the general form of the one generation Standard Model and show that most features (such as parity violation) are required by Left-handed Extended Lorentz group covariance. We will show the ElectroWeak symmetry group is $SU(2) \otimes U(1)$, and SU(2) doublets are left-handed and singlets are right-handed.

Thus much of the form of the Standard Model, which is often viewed as strange and unattractive, follows directly from (broken) Left-handed Extended Lorentz group covariance.

Hitherto, no one has satisfactorily explained (derived) the distinction that Nature makes between left-handedness and right-handedness. The breakdown of discrete symmetries has remained a mystery despite fifty years of research since their experimental discovery. The reaction of most theorists in this area is to view the Standard Model as a provisional theory primarily because of its "deformities." It does not have the appealing simplicity of Quantum Electrodynamics. Thus numerous attempts have been made to embed the Standard Model in theories with larger symmetries, and then to posit that the Standard Model emerges due to the breakdown of these larger theories. These approaches generally require the existence of copious numbers of new particles that have not been found experimentally.

We followed an alternative approach in deriving the left-right asymmetry of the Standard Model. *We establish the left-right asymmetry, not within the framework of a larger internal symmetry group or a larger dimensional space-time, but within an extension of the Lorentz group. This extension makes it reasonable to view neutrinos, and d, s, and b quarks, as tachyons.*[*] In addition to establishing a fundamental distinction between left-handedness and right-handedness (favoring left-handedness), it leads to the general form of the Standard Model. The Standard Model is not a malformed, hodge-podge theory (or a twisted, residue of a larger symmetric theory.)

---

[*] Since quarks are confined, the tachyon nature of d, s, and b quarks would not be directly seen, but would be reflected in quark momentum and spin distributions in baryons—currently an experimental area with many puzzling features that are not consistent with conventional parton, or other, models. The choice of d, s, and b quarks as tachyons is provisional; the u, c, and t quarks could well be the tachyons within the weak left-handed doublets. Experiment will determine which set of three are the tachyons.

To my wife, Margaret
With Love

CONTENTS

TABLES & FIGURES

0. Tachyons Exist!

In this prefatory chapter we establish the reality of tachyons with certitude inside Black Holes and show that many currently accepted theories contain tachyonic particles. Knowing that tachyons exist we can be confident that an acceptable formulation of tachyon quantum field theory must also exist.

0.1 Black Hole Tachyons

The conventional Schwarzschild solution of General Relativity has the metric

$$d\tau^2 = U(r)dt^2 - U(r)^{-1}dr^2 - r^2(d\theta^2 + \sin^2\theta d\varphi^2) \qquad (0.1)$$

It describes a non-rotating Black Hole if the mass M is within the Schwarzschild radius 2MG. The signs of the time and radial parts of the metric are reversed in the interior of a Black Hole where

$$U(r) = (1 - 2MG/r) < 0 \qquad (0.2)$$

.

We can establish a local inertial frame[1] at some point within a Black Hole and consider particle propagation within that reference frame. Particles moving only radially within that frame are tachyons. We see this when we consider the momentum of a particle of mass m. Its energy and radial momentum are[2]

$$p_0 = mE \qquad\qquad p^0 = mU^{-1}E \qquad (0.3)$$

and

---

[1] It is possible to establish a local inertial frame inside a Black Hole with a congruence since the number of positive signs in the Schwarzschild metric in the Black hole is the same as that of the Minkowski metric and the number of negative signs in the Schwarzschild metric is also the same as that of the Minkowski metric. See Turnbull (1961).

[2] See Blaha (2004) p. 71.

$$p_r = -mU^{-1}dr/d\sigma \qquad p^r = m\, dr/d\sigma \tag{0.4}$$

where σ is the proper time parameter. The mass condition for strictly radial motion is

$$p^2 = p_0p^0 + p_rp^r = m^2E^2U^{-1} - m^2U^{-1}(dr/d\sigma)^2 = m^2 \tag{0.5}$$

At any radial distance $r = r_a$ within the Schwarzschild radius, $U(r_a) < 0$ and thus a radially moving particle is a tachyon:

$$p^2 = m^2U^{-1}(E^2 - (dr/dp)^2) < 0 \tag{0.6}$$

Since there are many enormous Black Holes in the universe it is clear that tachyons exist in the universe in great quantity. Thus our development of the theory of faster-than-light motion and tachyons is justified, and not merely hypothetical.

0.2 Higgs Particles are Tachyonic

Higgs particles play an important role in Standard Model formulations. They implement spontaneous symmetry breaking and give masses to fermions and vector bosons. Higgs particles have "tachyonic" mass terms – m^2 term with the 'wrong" sign. They do have a quartic interaction term that enables a stable vacuum state to be defined. Yet they are tachyonic and show that m^2 terms with the 'wrong" sign appear in Nature if the Higgs mechanism of the Standard Model is correct.

0.3 Tachyons Appear in Many Superstring Theories

There are many SuperString theories that contain tachyons. Typically the definition of the physical particle states of a SuperString theory exclude states containing tachyons. Nevertheless the presence of tachyon fields within a SuperString theory requires that a proper formulation of quantum tachyon fields exist.

1. Why go Beyond the Light Barrier?

1.1 Einstein's Argument against Faster-than-Light Particles

Einstein considered the possibility of particles traveling faster than the speed of light. He dismissd this possibility based on the simple argument that the energy of a particle would approach infinity if the speed of the particle approached the speed of light:

$$E = m/(1 - v^2)^{½}$$

where we let the speed of light c = 1. Thus

$$E \rightarrow \infty \quad \text{as} \quad v \rightarrow 1$$

However, if particles were such that v > 1 when they were created or experienced a quantum jump (perhaps in a particle interaction) from v < 1 to v > 1 then faster than light particles might exist.

Based on this reasoning a number of authors[3] explored the possibility of formulating a quantum field theory of a faster than light particle for which Feinberg coined the word "tachyon." These authors assumed conventional Lorentz invariance and as a result could only consider spin 0 particles.[4] They were unable to define a local canonical free tachyon quantum field theory. So the question remained open until the first edition of this book[5] where it was resolved.

---

[3] S. Tanaka, Prog. Theoret. Phys. (Kyoto) **24**, 171 (1960); O. M. P. Bilaniuk, V. K. Deshpanda and E. C. G.; Sudarshan, Am J. Phys. **30**, 718 (1962); G. Feinberg, Phys. Rev. **159**, 1089 (1967); M. E. Arons & E. C. G. Sudarshan, Phys. Rev. **173**, 1622 (1968).

[4] E. P. Wigner, Ann. Math. **40**, 149 (1939); Yu. M. Shirokov, Soviet Phys-JETP **6**, 664, 919, 929 (1958); ___ 7, 493 (1958); ___ Nuc. Phys. **15**, 1, 13 (1960).

[5] S. Blaha, *A Derivation of ElectroWeak Theory based on an Extension of Special Relativity; Black Hole Tachyons; & Tachyons of Any Spin.* (Pingree-Hill Publishing, Auburn, NH, 2006) which will be called the first edition.

1.2 Why Transformations Between Reference Frames with a Relative Speed Greater Than Light Speed?

Earlier attempts at a tachyon theory were based solely on the Lorentz group, which consists of transformations between reference frames whose relative speed is less than c.

If we imagine measuring the coordinates of an event in a reference frame on a sub-light particle, and also imagine measuring the coordinates of an event in a reference frame on a tachyon, then it is reasonable to expect that the coordinates of the event in each frame are mathematically related to each other by some faster-than-light transformation.

Another situation which calls for transformations between reference frames with relative velocity greater than the speed of light is the case of two tachyons, for example one tachyon moving at 3c (three times the speed of light) and another tachyon moving at 7c (seven times the speed of light). Again it seems physically required that a faster than-light-transformation exists relating the coordinates of superluminal[6] reference frames residing on each tachyon.

The Lorentz group (actually the inhomogeneous Lorentz group) relates event coordinates for "normal" sub-light reference frames (i.e. the set of reference frames whose relative velocities have a magnitude less than c).

We will show that a similar set[7] of transformations relate a sub-light and superluminal reference frame, and also relate two "superluminal" reference frames[8] whose relative speed is greater than the speed of light. In the next chapter we will see that there are two possible groups of these types of transformations. We will choose one of these groups, which we call the Left-handed extended Lorentz group, based on a derivation of features of the Standard Model that follow from it. The other group leads to a **non**-Standard Model that favors right-handed features such as right-handed doublets.

---

[6] A superluminal reference frame is a reference frame moving faster than the speed of light relative to another reference frame. Since the distinction is one of *relative* speeds a superluminal reference frame from the viewpoint of one observer may be a "normal" reference frame the viewpoint of another observer. For example suppose we have three reference frames. In one reference frame an observer, observer1, sees the other two frames moving at a speed of 3.5c and 4c respectively in the positive x direction. Both of these frames appear to be superluminal to observer1. However an observer, observer2, on the 3.5c reference frame will see observer1's frame as superluminal and the 4c reference frame as a "normal" reference frame. A third observer, observer3, on the 4c reference frame will see observer1's frame as superluminal and the 3.5c reference frame as a "normal" reference frame. *Thus the concept of superluminal reference frame is relative.*

[7] This possibility was considered in the first edition. In this edition we have changed the coordinate transformation group to a different group: one of two possible groups.

[8] Reference frames that appear to be moving faster than the speed of light in a third reference frame.

1.3 The Set of Reference Frames

Under Special Relativity the set of Lorentz reference frames consists of all reference frames that are related by a Lorentz transformation where the magnitude of the relative velocity is less than the speed of light.[9]

If we extend Special Relativity to include transformations between reference frames where the magnitude of the relative velocity is not limited by the speed of light, then we obtain the set of reference frames for the extended Lorentz groups that will be described in this book. We will consider specific examples of this larger set of reference frames later.

1.4 Impact on General Relativity

The larger set of reference frames allowed by our extended Lorentz group will not change most of the basic tenets of General Relativity with several exceptions: 1) the concept of a locally inertial coordinate system[10] that appears in the equivalence principle is broadened to include the larger set of coordinate systems of the extended Lorentz group;[11] and 2) coordinate transformations can be complex and not congruences with important consequences for the field equations. Conventional General Relativity remains a subsector of this wider formulation. We will consider extended General Relativity in detail in a future work.

---

[9] Some efforts extend the set of Lorentz reference frames to include "infinite momentum frames" where the magnitude of the momentum $\rightarrow \infty$ as well. See for example S. Weinberg, Phys. Rev. **150**, 1313 (1966).

[10] A Lorentz, or extended Lorentz, reference frame is a locally inertial coordinate system.

[11] Since this larger set of locally inertial coordinate systems it is natural to consider the possibility of complex extensions of General Relativity such as in Blaha(2004).

Appendix 1-A. Phenomena Beyond the Light Barrier

1-A.1 Superluminal (Faster-than-Light) Transformations

In this Appendix we will briefly survey some of the very different features of faster-than light physical phenomena. We will frame our discussion in terms of the two simple reference frames depicted in Fig. 1-A.1. The prime frame is moving at a speed v > c (the speed of light) in the positive x direction with respect to the unprimed reference frame.

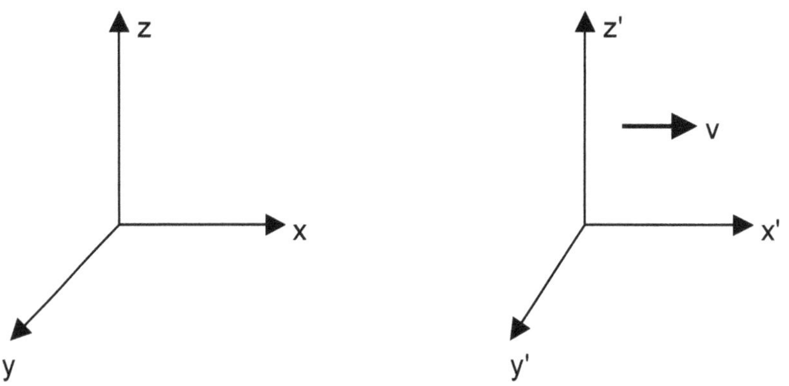

Figure 1-A.1. Two coordinate systems having a relative speed v in the x direction.

As shown later in the text (eq. 2.16 later) we define a superluminal (faster-than-light) transformation between coordinates in these reference frames with

$$t' = \gamma_s(t - \beta x/c)$$

$$x' = \gamma_s(x - \beta ct) \qquad (2.16)$$
$$y' = iy$$
$$z' = iz$$

where

$$\gamma_s = (\beta^2 - 1)^{-1/2} \qquad (2.13)$$

and $\beta = v/c > 1$. The appearance of imaginary values for y' and z' is not a cause for alarm. An observer resident in the prime coordinate system will measure real y and z distances with a ruler. The only purpose of the factors of i is to relate the y and z coordinates to y' and z'. An observer in either coordinate system will view his/her coordinates as real.

The energy and momentum of a tachyon (faster-than-light) particle of mass m traveling at a speed $v > c$ is

$$E = \gamma_s mc^2 \qquad (1\text{-A.1})$$

and

$$\mathbf{p} = m\gamma_s \mathbf{v} \qquad (1\text{-A.2})$$

Note that the tachyon defining condition is satisfied:

$$E^2 - c^2\mathbf{p}^2 = -m^2c^4 \qquad (1\text{-A.3})$$

Also note that as $\beta \rightarrow \infty$

$$E = 0 \qquad (1\text{-A.4})$$

and

$$p = mc \qquad (1\text{-A.5})$$

where $p = |\mathbf{p}|$. Tachyons are always in motion. The minimal momentum of a tachyon is given by eq. 1-A.5. It corresponds to zero energy. It is the tachyon equivalent of Einstein's famous $E = mc^2$.

1-A.2 Length Dilations and Time Contractions

In ordinary Lorentz transformations a moving ruler will appear to be shorter in the direction of its motion when measured in another reference frame. This phenomenon is called *Lorentz contraction.*

Superluminal Length Dilation/Contraction

In the case of a superluminal transformation we find precisely the opposite effect, *superluminal length dilation*, is a possibility. Consider the case of the transformation of eq. 2.16 above (coresponding to Fig. 1-A.1), which relates the prime reference frame traveling at speed v in the positive x direction to the unprimed reference frame. A ruler perpendicular to the x-axis will have the same length in both reference frames if its endpoints are simultaneously measured – perhaps by photographing it. The y and z equations in eqs. 2.16 specify this fact up to an extraneous factor of i.

If the ruler is at rest in the prime reference frame and parallel to the x' axis, then a simultaneous measurement of its endpoints at the same time t_0 by an observer in the unprimed reference frame (perhaps by photographing it) will reveal both *length contraction and dilation* depending on the value of β. If the length is $L' = x'_2 - x'_1$ in the prime frame and $L = x_2 - x_1$ in the unprimed frame, then the equations:

$$x'_1 = \gamma_s(x_1 - \beta c t_0) \qquad (1\text{-A.7})$$
$$x'_2 = \gamma_s(x_2 - \beta c t_0) \qquad (1\text{-A.8})$$

imply

$$L' = \gamma_s L = (\beta^2 - 1)^{-1/2} L \qquad (1\text{-A.9})$$

Thus we have three cases:

Case 1: $\beta \in <1, \sqrt{2}>$:	$L < L'$	Contraction	(1-A.10)
Case 2: $\beta = \sqrt{2}$:	$L = L'$	Equality	(1-A.11)
Case 3: $\beta \in <\sqrt{2}, \infty>$:	$L > L'$	Dilation	(1-A.12)

Superluminal Time Contraction/Dilation

In the case of a superluminal transformation we find *superluminal time contraction* is a possibility. Consider again the case of the transformation of eq. 2.16 above coresponding to Fig. 1-A.1 relating the prime reference frame traveling at speed v in the positive x direction to the unprimed reference frame. Consider the time interval between two events occurring at the same point x'_0 in the prime reference frame. From the viewpoint of an observer in the unprimed frame the events take place at different points x_1 and x_2. If the time interval is $T' = t'_2 - t'_1$ in the prime frame and $T = t_2 - t_1$ in the unprimed frame, then the inverse of eqs. 2.16 give:

$$t_1 = \gamma_s(t'_1 + \beta x'_0/c) \quad (1\text{-A}.13)$$
$$t_2 = \gamma_s(t'_2 + \beta x'_0/c) \quad (1\text{-A}.14)$$

and imply

$$T = \gamma_s T' = (\beta^2 - 1)^{-1/2} T' \quad (1\text{-A}.15)$$

Again we have three cases:

Case 1: $\beta \in <1, \sqrt{2}>$: $T > T'$ Dilation (1-A.16)

Case 2: $\beta = \sqrt{2}$: $T = T'$ Equality (1-A.17)

Case 3: $\beta \in <\sqrt{2}, \infty>$: $T < T'$ Contraction (1-A.18)

The time interval in the unprimed frame can be less than, equal to, or greater than the time interval in the frame where the events take place at the same spatial point.

Thus superluminal transformations are more complex than Lorentz transformations with respect to space and time, dilation and contraction.

1-A.3 Tachyon Fission to More Massive Particles

Another way in which faster-than-light phenomena differ from sublight phenomena is particle fission. Normally when a particle or nucleus decays or fissions the masses of the particles produced by the decay is smaller than the mass of the original particle or nucleus. And energy is released. We are familiar with fission as the source of nuclear energy.

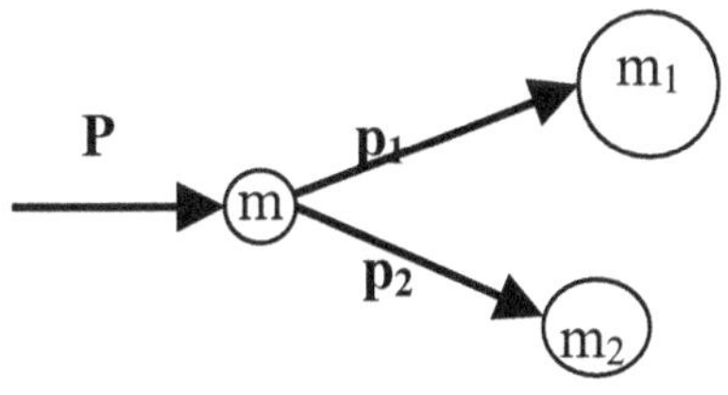

Figure 1-A.2. Two particle decay of a tachyon.

In the case of faster-than-light particles, tachyons, a much different possibility is present: a tachyon can decay into heavier tachyons. We will consider the specific case of a tachyon decaying into two particles to illustrate this possibility. (See Fig. 1-A.2.)

We will assume the initial tachyon has zero energy and thus the tachyons emerging from the decay also have zero energy. The analysis is based on conservation of total energy and momentum. We also assume tachyon particles have non-negative energies.

Momentum conservation implies

$$\mathbf{P} = \mathbf{p}_1 + \mathbf{p}_2 \tag{1-A.19}$$

Since all energies are zero

$$(cP)^2 = (c\mathbf{P})^2 = m^2$$

$$(cp_1)^2 = (c\mathbf{p}_1)^2 = m_1^2 \tag{1-A.20}$$

$$(cp_2)^2 = (c\mathbf{p}_2)^2 = m_2^2$$

where $P = |\mathbf{P}|$, $p_1 = |\mathbf{p}_1|$, and $p_2 = |\mathbf{p}_2|$. If we now square eq. 1-A.19 and use eqs. 1-A.20 we obtain

$$m^2 = m_1^2 + m_2^2 + 2m_1m_2 \cos\theta \tag{1-A.21}$$

where θ is the angle between the emerging particles momenta $\mathbf{p}_1$ and $\mathbf{p}_2$.

Eq. 1-A.21 has a number of interesting cases:

Case $\theta = 0$:

$$m = m_1 + m_2 \tag{1-A.22}$$

The masses of the outgoing tachyons sum to the mass of the original tachyon.

Case $\theta = \pi/2$:

$$m^2 = m_1^2 + m_2^2 \tag{1-A.23}$$

The masses of each outgoing tachyon is less than the mass of the original tachyon.

Case $\theta = \pi$:

$$m^2 = (m_1 - m_2)^2 \tag{1-A.24}$$

In this case either $m_1 > m$ or $m_2 > m$. Thus one of the outgoing tachyons has a greater mass than the original tachyon. Mass is effectively created from the spatial momentum of the particle. This process is the inverse of normal particle decay or fission where the sum of the outgoing masses is always less than the original particle's mass and the difference is mass converted into energy in the form of additional photons via "$E = mc^2$".

This last case, where one of the outgoing particles is more massive than the original particle, is not just for $\theta = \pi$. Since

$$\cos\theta = (m^2 - m_1^2 - m_2^2)/(2m_1m_2) \tag{1-A.25}$$

we see that *the sum of the outgoing tachyon masses is always greater than the original tachyon mass (except when $\theta = 0$)* since

$$\cos\theta = 1 + [m^2 - (m_1 + m_2)^2]/(2m_1m_2) \le 1 \tag{1-A.26}$$

and thus

$$[m^2 - (m_1 + m_2)^2]/(2m_1m_2) \le 0 \tag{1-A.27}$$

Note $m = m_1 + m_2$ only if $\theta = 0$.

Since we can transform the above discussion to the case of tachyons with a non-zero energy using an ordinary Lorentz transformation the above discussion in this subsection is general.

We therefore conclude that when a tachyon decays into two tachyons the sum of the masses of the produced tachyons is greater than the mass of the original tachyon except if the angle between the momenta of the produced tachyons is zero. In that case the sum of the masses of the produced tachyon equals the mass of the original tachyon.

1-A.4 Light Chasing Faster-than-Light Particles

Einstein told a story that he imagined positioning himself in a (Galilean) reference frame moving at the speed of light and seeing electromagnetic waves "frozen" in time so that they were no longer vibrating. This vision inspired him to reconsider the transformation laws between coordinate systems and to derive the theory of Special Relativity. In Special Relativity the speed of light is the same in all reference frames.

In this subsection we will consider a light pulse from the points of view of two reference frames whose relative speed v is greater than the speed of light. We will use the example considered earlier and add a pulse of light traveling in the positive x direction. (See Fig. 1-A.3.)

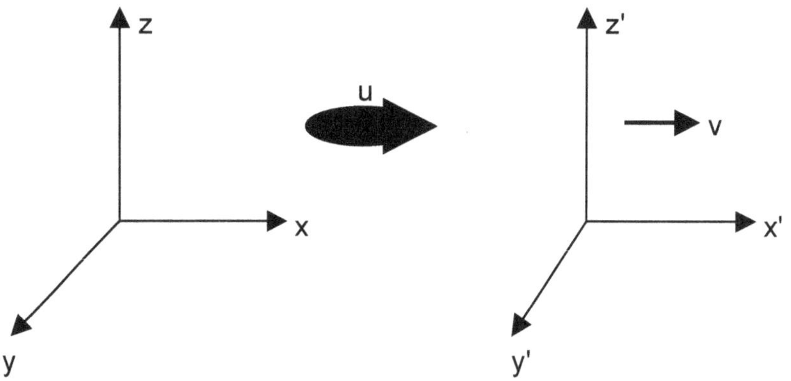

Figure 1-A.3. Two coordinate systems having a relative speed v in the x direction. A pulse of light is displayed as a thick arrow.

Eq. 2.17 (derived later) contains the general law for the addition of velocities in a situation such as depicted in Fig. 1-A.3. If we adapt it to the present example and let u be the speed of the pulse in the unprimed frame (temporarily forgetting it is a light pulse) we find eq. 2.17 implies

$$u' = (u - \beta c)/(1 - \beta u/c) \qquad (2.17a)$$

where $\beta = v/c > 1$, and u' is the speed of the pulse in the prime frame. Then if we set u = c we see that u' = c as well. *Thus our superluminal transformations preserve the constancy of the speed of light just like Lorentz transformations.*

As a result the pulse of light will intersect the z' axis eventually. However if superluminal transformations did not preserve the speed of light in all frames the pulse might never reach the z' axis. For example under a Galilean transformation the speed of the pulse would be u' = u – v = c – v and the pulse would actually be falling further and further behind the z' axis.

1-A.5 Electromagnetic Field of a Charged Tachyon – A Pancake Effect?

The electric field of a charge q at rest in a reference frame is:

$$\mathbf{E} = (q/(4\pi\varepsilon_0)\check{\mathbf{r}}/r^2 \tag{1-A.28}$$

in spherical coordinates where $\check{\mathbf{r}}$ is a unit vector in the radial direction.

Sublight Charged Particle

The electric and magnetic fields of a charge q moving in the positive x direction with speed v < c are

$$\mathbf{E} = (q/(4\pi\varepsilon_0)\check{\mathbf{r}}(1 - \beta^2)/[r^2(1 - \beta^2\sin^2\theta)^{\frac{3}{2}}] \tag{1-A.29}$$

$$\mathbf{B} = (q/(4\pi\varepsilon_0)\check{\mathbf{r}}\beta(1 - \beta^2)\sin\theta/[r^2(1 - \beta^2\sin^2\theta)^{\frac{3}{2}}] \tag{1-A.30}$$

where $\check{\mathbf{r}}$ is the radial unit vector, $\beta = v/c$, and θ is measured with respect to the polar axis which is taken to be the x axis. As $\beta \rightarrow 1$ the electric and magnetic fields develop a "pancake" form with large field strengths in the directions perpendicular to the direction of motion similar to the transverse fields of electromagnetic quanta. This feature is the basis of the Weizsäcker-Williams method of virtual quanta.

Charged Tachyon

The electric and magnetic fields of a tachyon of charge q moving in the positive x direction with speed v > c are

$$\mathbf{E} = (q/(4\pi\varepsilon_0)\check{\mathbf{r}}(\beta^2 - 1)/[r^2(\beta^2\sin^2\theta - 1)^{\frac{3}{2}}] \tag{1-A.31}$$

$$\mathbf{B} = (q/(4\pi\varepsilon_0)\check{\mathbf{r}}\beta(\beta^2 - 1)\sin\theta/[r^2(\beta^2\sin^2\theta - 1)^{\frac{3}{2}}] \tag{1-A.32}$$

where $\beta = v/c > 1$, and θ is again measured with respect to the polar axis which is taken to be the x axis. In the case of tachyons there are three cases of interest.

Case $\beta^2\sin^2\theta - 1 < 0$:

The electric and magnetic fields are pure imaginary and are excluded from the forward and backward cones surrounding the x axis defined by $|\sin\theta| < \beta^{-1}$.

Case $\beta^2\sin^2\theta - 1 = 0$:

The electric and magnetic fields are infinite. *Thus the field strengths are infinite on a cone at the angle θ with respect to the x-axis. By comparison, a magnetic monopole only has a one-dimensional, singularity line extending from the monopole to infinity.*

Case $\beta^2\sin^2\theta - 1 > 0$:

The electric and magnetic fields decrease in strength as $\sin^2\theta$ increases. Thus the region of maximum field strength are the forward and backward cones where $|\sin\theta|$ is greater than but near β^{-1} in value. The pancake picture of the sublight charged particle does not hold for charged tachyons.

1-A.6 Superluminal (Tachyon) Physics is Different

The simple classical examples presented in this appendix demonstrate that superluminal physics has many interesting new features that are worthy of interest. Since tachyons exist in Black Holes, and, perhaps, in other contexts, their study is a worthwhile endeavor.

2. Extended Lorentz Groups

2.1 The Lorentz Group

The Lorentz group, with which we are familiar, relates the coordinates of an event in two coordinate systems that differ by a relative velocity whose magnitude is less than the speed of light. The inhomogenous Lorentz group includes coordinate displacements. We will discuss the relationship between coordinates modulo displacements, and not consider the displacement generators, since they will be the same for both ordinary Lorentz group transformations, and the transformation groups for the case of relative velocities of any magnitude. Thus we will consider the homogeneous Lorentz group, and its generalizations that include faster-than-light transformations.

The elements of the Lorentz group, $\Lambda(\mathbf{v})$, when treated as 4×4 matrices satisfy

$$\Lambda(\mathbf{v})^{T} G \Lambda(\mathbf{v}) = G \tag{2.1}$$

where G is the 4 x 4 diagonal matrix form of the metric diag(1, – 1, –1, –1), the superscript "T" indicates the transpose, and **v** is the relative velocity of the reference frames. Note g^{00} is +1 in our metric. When $\Lambda(\mathbf{v})$ is a boost the components[12] of its matrix representation are real:

$$\Lambda(\mathbf{v}) = \begin{bmatrix} \gamma & -\gamma v_x & -\gamma v_y & -\gamma v_z \\ -\gamma v_x & 1 + (\gamma - 1)v_x^2/v^2 & (\gamma - 1)v_x v_y/v^2 & (\gamma - 1)v_x v_z/v^2 \\ -\gamma v_y & (\gamma - 1)v_x v_y/v^2 & 1 + (\gamma - 1)v_y^2/v^2 & (\gamma - 1)v_y v_z/v^2 \\ -\gamma v_z & (\gamma - 1)v_x v_z/v^2 & (\gamma - 1)v_y v_z/v^2 & 1 + (\gamma - 1)v_z^2/v^2 \end{bmatrix} \tag{2.2}$$

[12] We shall consider only the proper, orthochronous Lorentz group until section 2.3 where we extend the discussion to the other subgroups of the Lorentz group and extended Lorentz groups.

where $\gamma = (1 - v^2)^{-1/2}$, $c = 1$, $\mathbf{v} = (v_x, v_y, v_z)$, and $v = |\mathbf{v}|$. The boost transformation $\Lambda(\mathbf{v})$ can be expressed in the form

$$\Lambda(\mathbf{v}) = \exp[i\omega\hat{\mathbf{u}}\cdot\mathbf{K}] \tag{2.3}$$

where $\mathbf{v} = \hat{\mathbf{u}}\ \tanh\omega$, $\hat{\mathbf{u}}\bullet\hat{\mathbf{u}} = 1$, and $\mathbf{K}$ is the boost vector. Using the unit normalized velocity vector $\mathbf{u} = (u_x, u_y, u_z)$ the Lorentz boost matrix can be written:

$$\Lambda(\omega, \mathbf{u}) = \Lambda(\mathbf{v}) \tag{2.4}$$

$$= \begin{bmatrix} \cosh(\omega) & -\sinh(\omega)u_x & -\sinh(\omega)u_y & -\sinh(\omega)u_z \\ -\sinh(\omega)u_x & 1 + (\cosh(\omega) - 1)u_x^2 & (\cosh(\omega) - 1)u_xu_y & (\cosh(\omega) - 1)u_xu_z \\ -\sinh(\omega)u_y & (\cosh(\omega) - 1)u_xu_y & 1 + (\cosh(\omega) - 1)u_y^2 & (\cosh(\omega) - 1)u_yu_z \\ -\sinh(\omega)u_z & (\cosh(\omega) - 1)u_xu_z & (\cosh(\omega) - 1)u_yu_z & 1 + (\cosh(\omega) - 1)u_z^2 \end{bmatrix}$$

This definition of the general form of the proper, orthochronous, Lorentz boost matrix $\Lambda(\omega, \mathbf{u})$ will be used in subsequent sections to define extended faster-than-light transformations.

The vector form of a Lorentz boost transformation is

$$\begin{aligned} \mathbf{x}' &= \mathbf{x} + (\gamma - 1)\mathbf{x}\cdot\mathbf{v}\ \mathbf{v}/v^2 - \gamma\mathbf{v}t \\ t' &= \gamma(t - \mathbf{v}\cdot\mathbf{x}/c^2) \end{aligned} \tag{2.5}$$

where $\gamma = (1 - \beta^2)^{-1/2}$ with $\beta = v/c = v$ (with $c = 1$).

2.2 Sets of Real Matrices $\Lambda(\omega, u)$ for Complex ω

Faster-than-light transformations will be implemented with specific complex forms of ω as we will see in the next section. Since

$$\cosh^2(z) - \sinh^2(z) = 1 \tag{2.6}$$

for complex values z it is easy to show that

$$\Lambda(\omega, \mathbf{u})^T G \Lambda(\omega, \mathbf{u}) = G \qquad (2.7)$$

for any complex ω. Eq. 2.7 implies $\Lambda(\omega, \mathbf{u})$ is a member of the Lorentz group or of the complex Lorentz group for complex ω.

For certain values of the imaginary part of ω the matrix $\Lambda(\omega, \mathbf{u})$ has a simple form, similar to that of $\Lambda(\omega, \mathbf{u})$ for real ω, but which generates boosts to relative speeds greater than the speed of light. Among these values are:

$$\omega \rightarrow \omega_{\pm} = \omega \pm i\pi/2 \qquad (2.8)$$

2.2 Extensions of the Lorentz Group to Faster-than-Light Transformations

In this section we will substitute $\omega_{\pm}$ for ω in $\Lambda(\omega, \mathbf{u})$ and then show that we obtain two sets of possible transformations to faster-than-light reference frames. One set of transformations where $\omega_L = \omega + i\pi/2$ will be called *left-handed superluminal transformations (boosts)*. When they are combined with the Lorentz group we are led to the "left-handed" Standard Model. We denote members of this set, $\Lambda_L(\omega, \mathbf{u})$, with the subscript "L".

The other set of transformations where $\omega_R = \omega - i\pi/2$ will be called *right-handed superluminal transformations*. When they are combined with the Lorentz group they lead to a right-handed, unphysical, version of the Standard Model. We denote members of this set, $\Lambda_R(\omega, \mathbf{u})$, with the subscript "R".

Before considering faster-than-light boosts we note the relation between ω in a Lorentz boost $\Lambda(\omega, \mathbf{u})$, and the magnitude of the relative velocity $v < 1$, is

$$\mathbf{v} = \hat{\mathbf{u}} \tanh\omega \qquad \hat{\mathbf{u}}\bullet\hat{\mathbf{u}} = 1$$

$$\cosh(\omega) = \gamma = (1 - v^2)^{-1/2} \qquad (2.9)$$

$$\sinh(\omega) = v\gamma = \beta\gamma$$

where $\beta = v = |\mathbf{v}|$.

Left-Handed Superluminal Transformations

Left-handed (proper orthochronous) superluminal boost transformations $\Lambda_L(\mathbf{v})$ have the same form as eq. 2.2 for ordinary (proper orthochronous) Lorentz boost transformations. However the magnitude of the relative velocity **v** is greater than the speed of light. Thus $\gamma = (1 - v^2)^{-½}$ is pure imaginary and $\Lambda_L(\mathbf{v})$ is complex.

$$\Lambda_L(\mathbf{v}) = \begin{bmatrix} \gamma & -\gamma v_x & -\gamma v_y & -\gamma v_z \\ -\gamma v_x & 1 + (\gamma - 1)v_x^2/v^2 & (\gamma - 1)v_x v_y/v^2 & (\gamma - 1)v_x v_z/v^2 \\ -\gamma v_y & (\gamma - 1)v_x v_y/v^2 & 1 + (\gamma - 1)v_y^2/v^2 & (\gamma - 1)v_y v_z/v^2 \\ -\gamma v_z & (\gamma - 1)v_x v_z/v^2 & (\gamma - 1)v_y v_z/v^2 & 1 + (\gamma - 1)v_z^2/v^2 \end{bmatrix} \qquad (2.10)$$

This transformation raises several issues – the most prominent of which is the interpretation of the imaginary coordinates generated by the transformation. Imaginary coordinates would appear at first glance to be unphysical. However if we view the measurement of these quantities operationally: an observer measures distances with "rulers", and time with clocks, which both give real numeric values. Thus an observer in a coordinate system with imaginary coordinates will be "unaware" of the imaginary nature of these quantities. It is only when the observer's coordinate system is related to another coordinate system through a superluminal transformation as done above that the imaginary nature of the coordinates becomes evident. From this point of view imaginary coordinates pose no new conceptual issues.

It will become apparent that a related transformation defined by

$$E(\mathbf{v}) = i\Lambda_L(\mathbf{v}) \qquad (2.10a)$$

should be used to define the coordinates generated by a superluminal boost transformation.[13] This definition of coordinates leads to the tachyonic Dirac equation for spin ½ tachyons. Therefore we define

$$X' = E(\mathbf{v})X = i\Lambda_L(\mathbf{v})X \qquad (2.10b)$$

---

[13] E is an upper case epsilon chosen to stand for Extended since the boosts in question are extended boosts ($v > 1$) that are not part of the Lorentz group.

where X' and X are coordinates in the prime and unprimed reference frames respectively. X' and X are represented as column vectors.[14] If we consider the simple case of a relative velocity v in the x direction, then the E boost gives

$$t' = |\gamma|(t - \beta x)$$
$$x' = |\gamma|(x - \beta t)$$
$$y' = iy$$
$$z' = iz$$

where $|\gamma|$ is the absolute value of γ. Thus y' and z' are imaginary from the viewpoint of the unprimed coordinate system.

Eqs. 2.7 and 2.10a imply

$$E(\omega, \mathbf{u})^T G E(\omega, \mathbf{u}) = -G \qquad (2.10c)$$

where $E(\omega, \mathbf{u}) \equiv E(\mathbf{v})$. Thus E inverts the sign of the metric tensor. As a result the invariant scalar product relation (in matrix form) is

$$XGY = X'(-G)Y' \qquad (2.10d)$$

The metric tensor is represented by G. Most of the succeeding discussion will be phrased in terms of $\Lambda_L(\mathbf{v})$. Eq. 2.10a enables one to easily rephrase the discussion in terms of $E(\omega, \mathbf{u})$.

The Cosh-Sinh Representation of Left-Handed Superluminal Boosts

We will now develop the representation of left-handed superluminal boost transformations in terms of cosh(ω) and sinh(ω) for later use in our discussion of tachyons. We find that we must use a complex $\omega = \omega_L = \omega + i\pi/2$ to properly describe left-handed superluminal boosts. The relation between ω_L and v is different from eq. 2.9 for the case of left-handed superluminal boosts:

$$\cosh(\omega_L) = i \sinh(\omega) = -\gamma = i\,\gamma_s$$
$$\sinh(\omega_L) = i \cosh(\omega) = -\beta\gamma = i\beta\,\gamma_s \qquad (2.11)$$

---

[14] This definition of superluminal coordinate transformations is physically acceptable as long as allowance is made for the change in the definition of the invariant interval based on eq. 2.10c.

where $\beta = v > 1$ and $\boldsymbol{\omega} \geq \mathbf{0}$. Note

$$\sinh(\omega) = \gamma_s$$
$$\cosh(\omega) = \beta\gamma_s \qquad (2.12)$$

with

$$\gamma_s = (\beta^2 - 1)^{-\frac{1}{2}} \qquad (2.13)$$

Upon substituting ω_L for ω in eq. 2.4 we obtain another form (equivalent to that of eq. 2.10) for a left-handed superluminal transformation:

$$\Lambda_L(\omega, \mathbf{u}) = \Lambda(\omega + i\pi/2, \mathbf{u}) = -iE(\omega, \mathbf{u})$$

$$= \begin{bmatrix} \cosh(\omega_L) & -\sinh(\omega_L)u_x & -\sinh(\omega_L)u_y & -\sinh(\omega_L)u_z \\ -\sinh(\omega_L)u_x & 1+(\cosh(\omega_L)-1)u_x^2 & (\cosh(\omega_L)-1)u_xu_y & (\cosh(\omega_L)-1)u_xu_z \\ -\sinh(\omega_L)u_y & (\cosh(\omega_L)-1)u_xu_y & 1+(\cosh(\omega_L)-1)u_y^2 & (\cosh(\omega_L)-1)u_yu_z \\ -\sinh(\omega_L)u_z & (\cosh(\omega_L)-1)u_xu_z & (\cosh(\omega_L)-1)u_yu_z & 1+(\cosh(\omega_L)-1)u_z^2 \end{bmatrix}$$

$$= \begin{bmatrix} i\gamma_s & -i\beta\gamma_su_x & -i\beta\gamma_su_y & -i\beta\gamma_su_z \\ -i\beta\gamma_su_x & 1+(i\gamma_s-1)u_x^2 & (i\gamma_s-1)u_xu_y & (i\gamma_s-1)u_xu_z \\ -i\beta\gamma_su_y & (i\gamma_s-1)u_xu_y & 1+(i\gamma_s-1)u_y^2 & (i\gamma_s-1)u_yu_z \\ -i\beta\gamma_su_z & (i\gamma_s-1)u_xu_z & (i\gamma_s-1)u_yu_z & 1+(i\gamma_s-1)u_z^2 \end{bmatrix} = \Lambda_L(\mathbf{v}) \qquad (2.14)$$

by eq. 2.10.

A simple case that illustrates the nature of the left-handed superluminal boost is to assume the relative velocity is in the x direction. Then eq. 2.14 becomes

$$\Lambda_L(\omega, \mathbf{u} = (1,0,0)) = \begin{bmatrix} i\gamma_s & -i\beta\gamma_s & 0 & 0 \\ -i\beta\gamma_s & i\gamma_s & 0 & 0 \\ 0 & 0 & 1 & 0 \\ 0 & 0 & 0 & 1 \end{bmatrix} \qquad (2.15)$$

implementing the coordinate transformation:

$$X' = -i\Lambda_L(\omega, \mathbf{u} = (1,0,0))X = E(\omega, \mathbf{u} = (1,0,0))X$$

or

$$\begin{aligned} t' &= \gamma_s(t - \beta x) \\ x' &= \gamma_s(x - \beta t) \\ y' &= iy \\ z' &= iz \end{aligned} \tag{2.16}$$

The addition rule for the x-component of velocity can be computed for infinitesimal displacements in space and time:

$$v_x' = \Delta x' / \Delta t' = (\Delta x\, \gamma_s - \Delta t\, \beta\gamma_s)/(\Delta t\, \gamma_s - \Delta x\, \beta\gamma_s)$$

$$= (v_x - \beta)/(1 - \beta v_x) \tag{2.17}$$

(where $\beta = u$ is the relative speed) in the limit $\Delta t \rightarrow 0$ where the x component of a particle's velocity in the unprimed frame is $v_x = \Delta x/\Delta t$. $\Delta t'$ is determined by

$$\Delta t' = \Delta t\, \gamma_s(1 - \beta v_x) \tag{2.18}$$

Note the velocity of light is the same in the primed and unprimed reference frames. (If $v_x = 1$ then $v_x' = 1$.) *Thus left-handed superluminal transformations preserve the constancy of the speed of light in all reference frames.*

Further note that increasing the value of ω in $\Lambda_L(\omega, \mathbf{u})$ corresponds to decreasing the magnitude of the relative velocity v since

$$v = \text{cotanh}(\omega) \tag{2.19}$$

Thus when $v = 1$ then $\omega = \infty$, and when $\omega = 0$ then $v = \infty$.

General Velocity Transformation Law – Left-Handed Superluminal Boosts

The general velocity transformation law for a particle moving with velocity **v** in the unprimed reference frame and velocity **v'** in the primed reference frame is

$$\mathbf{v}' = \mathbf{w} + (\gamma - 1)\mathbf{w}\cdot\mathbf{v}\, \mathbf{w}/w^2 - \gamma\mathbf{w} \tag{2.20}$$

where **w** is the relative velocity of the primed reference frame with respect to the unprimed reference frame, and $\gamma = (1 - w^2)^{-1/2}$. Eq. 2.20 is obtained by calculating the derivative d**x**'/dt' using eqs. 2.5. The relative velocity **w** can be greater or less than the speed of light. Eq. 2.20 implies

$$v'^2 = 1 + (v^2 - 1)(1 - w^2)/(1 - \mathbf{w}\cdot\mathbf{v})^2 \qquad (2.21)$$

The velocity transformation law can be used to determine the multiplication rules for Lorentz and extended Lorentz transformations (next subsection).

Left-Handed Transformations Multiplication Rules

In this subsection we will determine the multiplication rules of left-handed extended Lorentz transformations and show the Lorentz group is an invariant subgroup of the extended left-handed Lorentz group. To do this we will consider three reference frames: an unprimed frame, a "primed" frame moving with velocity **w** with respect to the unprimed frame, and a "double-primed" frame moving with velocity **v** with respect to the unprimed frame and velocity **v'** with respect to the primed frame. See Fig. 2.1.

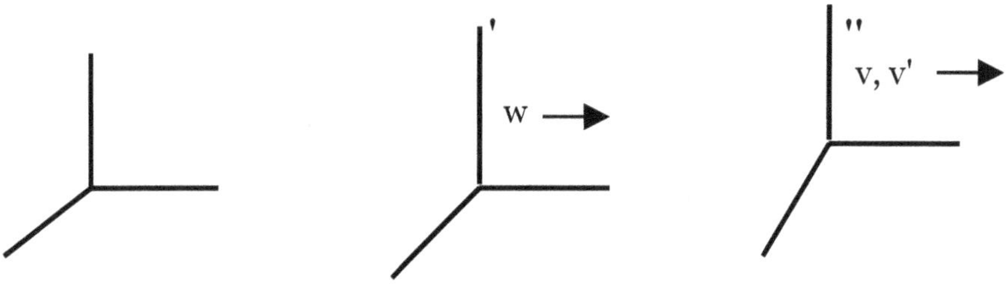

Figure 2.1. Three reference frames used to establish transformation multiplication rules.

A moment's consideration reveals that **v'** is related to **v** by eqs. 2.20 and 2.21. (Think of the double-primed coordinate system as a particle or attached to a particle. In addition note that the transformation law from the unprimed to the double-primed reference frame must be the product of consecutive transformations (boosts) from the unprimed to the primed reference frames and then from the primed to the double-primed reference frames:

$$\Lambda_?(\mathbf{v}) = \Lambda_?(\mathbf{v'})\Lambda_?(\mathbf{w}) \qquad (2.22)$$

where the "?" subscripts indicate Lorentz or superluminal transformations (boosts) depending on the magnitude of the relative velocity in the transformation's parentheses.

We now consider the cases using eq. 2.21:

1) If $w > 1$ and $v' > 1$

   then eq. 2.21 implies $v < 1$ and thus $\Lambda_?(\mathbf{v})$ is a Lorentz transformation

$$\Lambda(\mathbf{v}) = \Lambda_L(\mathbf{v}')\Lambda_L(\mathbf{w}) \tag{2.23}$$

2) If $w > 1$, $v' < 1$

   then eq. 2.21 implies $v > 1$ and thus $\Lambda_?(\mathbf{v})$ is a superluminal transformation

$$\Lambda_L(\mathbf{v}) = \Lambda(\mathbf{v}')\Lambda_L(\mathbf{w}) \tag{2.24}$$

3) If $w < 1$, $v' > 1$

   then eq. 2.21 implies $v > 1$ and thus $\Lambda_?(\mathbf{v})$ is a superluminal transformation

$$\Lambda_L(\mathbf{v}) = \Lambda_L(\mathbf{v}')\Lambda(\mathbf{w}) \tag{2.25}$$

4) If $w < 1$, $v' < 1$

   then eq. 2.21 implies $v < 1$ and thus $\Lambda_?(\mathbf{v})$ is a Lorentz transformation

$$\Lambda(\mathbf{v}) = \Lambda(\mathbf{v}')\Lambda(\mathbf{w}) \tag{2.26}$$

where, in each above case, the transformation on the left side of the equation may be a boost or a combination of a boost and a spatial rotation. Thus we have obtained the multiplication rules for left-handed extended Lorentz transformations.

The inverse of a Lorentz boost (eq. 2.3) is

$$\Lambda^{-1}(\omega, \hat{\mathbf{u}}) = \exp[-i\omega\hat{\mathbf{u}}\cdot\mathbf{K}] \tag{2.27}$$

where $\boldsymbol{\omega} \geq \mathbf{0}$. It shows the inverse is generated by letting $\omega \to -\omega$. Note that since $v = \tanh\omega$, the effect of $\omega \to -\omega$ is to let $v \to -v$. In the case of superluminal left-handed boosts, since

$$\Lambda_L(\omega, \mathbf{u}) = \Lambda(\omega + i\pi/2, \mathbf{u}) = \exp[i(\omega + i\pi/2)\hat{\mathbf{u}}\cdot\mathbf{K}] \tag{2.28}$$

we find the inverse is

$$\Lambda_L^{-1}(\omega, \mathbf{u}) = \Lambda(-(\omega + i\pi/2), \mathbf{u}) = \exp[-i(\omega + i\pi/2)\hat{\mathbf{u}}\cdot\mathbf{K}] \qquad (2.29)$$

where $\boldsymbol{\omega} \geq \mathbf{0}$. Since $\Lambda_L^{-1}(\omega, \mathbf{u})$ is not the hermitean conjugate of $\Lambda_L(\omega, \mathbf{u})$, superluminal boosts are not unitary. However unitarity is not required since even complex Lorentz group elements satisfy the defining relation of the Lorentz group (eqs. 2.1 and 2.7). The effect of letting $\omega_L = \omega + i\pi/2 \rightarrow -(\omega + i\pi/2)$ is to let $v \rightarrow -v$ since

$$\tanh(\omega_L) = \beta = v \qquad (2.30)$$

by eqs. 2.11.

We now turn to proving the Lorentz group is an invariant subgroup of the left-handed extended Lorentz group.

Theorem: The Lorentz group is an invariant subgroup of the left-handed extended Lorentz group.
Proof:

Consider the quantity

$$I = \Lambda_L(\omega_u, \mathbf{u})^{-1}\Lambda(\omega_v, \mathbf{v})\Lambda_L(\omega_u, \mathbf{u})$$

then

$$I = \Lambda_L(\omega_u, \mathbf{u})^{-1}\Lambda_L(\omega_z, \mathbf{z})$$

$$= \Lambda(\omega_w, \mathbf{w})$$

by eqs. 2.24 and 2.23 respectively for some ω_z, $\mathbf{z}$, ω_w and $\mathbf{w}$. Thus the Lorenz group is an invariant subgroup of the left-handed extended Lorentz group.

Right-Handed Superluminal Transformations

When we transform between reference frames using a right-handed[15] superluminal boost (where the magnitude of the relative velocity v is greater than c) the relation between ω and v also changes. The variable ω becomes $\omega_R = \omega - i\pi/2$ and

[15] We call these transformations right-handed because they lead eventually to an alternate Standard Model with right-handed SU(2) doublets and left-handed SU(2) singlets. This alternate right-handed Standard Model does not appear to correspond to experimental reality.

$$\cosh(\omega_R) = -i\sinh(\omega) = \gamma = -i\gamma_s \tag{2.31}$$
$$\sinh(\omega_R) = -i\cosh(\omega) = \beta\gamma = -i\beta\gamma_s \tag{2.32}$$

where $\beta = v > 1$ and $\boldsymbol{\omega} \geq \mathbf{0}$. Note (as before)

$$\sinh(\omega) = \gamma_s$$
$$\cosh(\omega) = \beta\gamma_s \tag{2.12}$$

with

$$\gamma_s = (\beta^2 - 1)^{-1/2} \tag{2.13}$$

Upon substituting ω_R for ω in eq. 2.4 we obtain the form of the right-handed superluminal transformation and a corresponding transformation E_R:

$$\Lambda_R(\omega, \mathbf{u}) = \Lambda(\omega - i\pi/2, \mathbf{u}) = -iE_R(\omega, \mathbf{u})$$

$$= \begin{bmatrix} -i\gamma_s & i\beta\gamma_s u_x & i\beta\gamma_s u_y & i\beta\gamma_s u_z \\ i\beta\gamma_s u_x & 1 + (-i\gamma_s - 1)u_x^2 & (-i\gamma_s - 1)u_x u_y & (-i\gamma_s - 1)u_x u_z \\ i\beta\gamma_s u_y & (-i\gamma_s - 1)u_x u_y & 1 + (-i\gamma_s - 1)u_y^2 & (-i\gamma_s - 1)u_y u_z \\ i\beta\gamma_s u_z & (-i\gamma_s - 1)u_x u_z & (-i\gamma_s - 1)u_y u_z & 1 + (-i\gamma_s - 1)u_z^2 \end{bmatrix} \tag{2.33}$$

A simple case that illustrates the nature of the right-handed superluminal transformation is to assume the relative velocity is in the x direction. Then eq. 2.33 becomes

$$\Lambda_R(\omega, \mathbf{u} = (1,0,0)) = \begin{bmatrix} -i\gamma_s & i\beta\gamma_s & 0 & 0 \\ i\beta\gamma_s & -i\gamma_s & 0 & 0 \\ 0 & 0 & 1 & 0 \\ 0 & 0 & 0 & 1 \end{bmatrix} \tag{2.34}$$

implementing the coordinate transformation:

$$X' = E_R(\omega, \mathbf{u})X = i\Lambda_R(\omega, \mathbf{u})X$$

or

$$t' = \gamma_s(t - \beta x)$$
$$x' = \gamma_s(x - \beta t) \tag{2.35}$$

$$y' = iy$$
$$z' = iz$$

Comparing eq. 2.34 with eq. 2.15 for a left-handed superluminal boost we see that

$$PT\Lambda_L(\omega, \mathbf{u} = (1,0,0)) = \begin{bmatrix} -i\gamma_s & i\beta\gamma_s & 0 & 0 \\ i\beta\gamma_s & -i\gamma_s & 0 & 0 \\ 0 & 0 & -1 & 0 \\ 0 & 0 & 0 & -1 \end{bmatrix} \qquad (2.36)$$

where P is the parity operator and T is the time reversal operator. If we now apply a spatial rotation $\mathfrak{R}$ of π radians around the x axis then we obtain

$$\mathfrak{R}PT\Lambda_L(\omega, \mathbf{u} = (1,0,0))\mathfrak{R}^{-1} = \begin{bmatrix} -i\gamma_s & i\beta\gamma_s & 0 & 0 \\ i\beta\gamma_s & -i\gamma_s & 0 & 0 \\ 0 & 0 & 1 & 0 \\ 0 & 0 & 0 & 1 \end{bmatrix} \qquad (2.37)$$

$$= \Lambda_R(\omega, \mathbf{u} = (1,0,0))$$

by eq. 2.34. Since P and T commute with spatial rotations we find

$$\Lambda_R(\omega, \mathbf{u} = (1,0,0)) = PT\mathfrak{R}\Lambda_L(\omega, \mathbf{u} = (1,0,0))\mathfrak{R}^{-1} \qquad (2.38)$$

or, more generally, performing additional spatial rotations:

$$\Lambda_R(\omega, \mathbf{u}) = PT\mathfrak{R}_u\mathfrak{R}\mathfrak{R}_w\Lambda_L(\omega, \mathbf{w})\mathfrak{R}_w^{-1}\mathfrak{R}^{-1}\mathfrak{R}_u^{-1} \qquad (2.39)$$

or,

$$\Lambda_R(\omega, \mathbf{u}) = PT\mathfrak{R}_{tot}\Lambda_L(\omega, \mathbf{w})\mathfrak{R}_{tot}^{-1} \qquad (2.40)$$

where **u** and **w** are unit vectors. Alternately,

$$\Lambda_L(\omega, \mathbf{w}) = PT\mathfrak{R}_{tot}^{-1}\Lambda_R(\omega, \mathbf{u})\mathfrak{R}_{tot} \tag{2.41a}$$

or

$$\Lambda_L(\omega, \mathbf{w}) = PT\Lambda_R(\omega, \mathbf{u}') \tag{2.41b}$$

for some unit vector $\mathbf{u}'$.

Thus we have shown that PT can be used to relate left-handed and right-handed boosts in a one-to-one fashion. *The appearance of the parity operator P takes on great significance when we derive features of the Standard Model. The appearance of left-handed doublets and right-handed singlets stems directly from the implicit parity dependence of the left-handed extended Lorentz group.*

For the right-handed boost of eq. 2.34 the addition rule for the x-component of velocity can be computed for infinitesimal displacements in space and time:

$$\begin{aligned} v_x' = \Delta x' / \Delta t' &= (\Delta x\, \gamma_s - \Delta t\, \beta\gamma_s)/(\Delta t\, \gamma_s - \Delta x\, \beta\gamma_s) \\ &= (v_x - \beta)/(1 - \beta v_x) \end{aligned} \tag{2.42}$$

in the limit $\Delta t \to 0$ where the x component of a particle's velocity in the unprimed frame is $v_x = \Delta x/\Delta t$. Note if $v_x = 1$ then $v_x' = 1$. *Thus right-handed superluminal transformations also preserve the constancy of the speed of light in all reference frames.*

2.3 Inhomogeneous Left-Handed Extended Lorentz Group

The *Left-Handed Extended Lorentz group* consists of the elements of the Lorentz group plus left-handed superluminal transformations including pure boosts and combinations of boosts and spatial rotations. Thus homogeneous left-handed superluminal transformations have the general form:

$$\Lambda_L(\mathbf{v}, \boldsymbol{\theta}) = \exp[i\omega_L\hat{\mathbf{u}}\cdot\mathbf{K} + i\boldsymbol{\theta}\cdot\mathbf{J}] \tag{2.43}$$

where $\omega_L' = \omega + i\pi/2$, $\boldsymbol{\theta}$ is the angular vector, and $\mathbf{J}$ is the angular momentum operator vector. Inhomogeneous left-handed superluminal transformations, which include displacements, can be expressed as

$$\Lambda_L(\mathbf{v}, \boldsymbol{\theta}, \mathbf{d}) = \exp[i\omega_L\hat{\mathbf{u}}\cdot\mathbf{K} + i\boldsymbol{\theta}\cdot\mathbf{J} - i\mathbf{d}\cdot\mathbf{P}] \tag{2.44}$$

where $\mathbf{P}$ is the momentum operator vector and $\mathbf{d}$ is a displacement vector.

We have verified the group structure of the left-handed extended Lorentz group is indeed a group in the preceding section (eqs. 2.23 – 2.29).

We also note

$$\det \Lambda_L(\omega, \mathbf{u}) = \pm 1 \tag{2.45}$$

by eq. 2.7.

The ordinary Lorentz group is divided into four disjoint subgroups that are often denoted:

$$\begin{aligned}
L_+^\uparrow&: \quad \det \Lambda(\omega, \mathbf{u}) = +1; \;\; \operatorname{sgn} \Lambda(\omega, \mathbf{u})^0{}_0 = +1 \\
L_-^\uparrow&: \quad \det \Lambda(\omega, \mathbf{u}) = -1; \;\; \operatorname{sgn} \Lambda(\omega, \mathbf{u})^0{}_0 = +1 \\
L_+^\downarrow&: \quad \det \Lambda(\omega, \mathbf{u}) = +1; \;\; \operatorname{sgn} \Lambda(\omega, \mathbf{u})^0{}_0 = -1 \\
L_-^\downarrow&: \quad \det \Lambda(\omega, \mathbf{u}) = -1; \;\; \operatorname{sgn} \Lambda(\omega, \mathbf{u})^0{}_0 = -1
\end{aligned} \tag{2.46}$$

where sgn $\Lambda(\omega, \mathbf{u})^0{}_0$ is the sign of the 00 component of the $\Lambda(\omega, \mathbf{u})$ matrix. The various subgroups are related by the discrete transformations of parity P and time reversal T:

$$\begin{aligned}
L_+^\uparrow &\xrightarrow{\text{P}} L_-^\uparrow \\
L_+^\uparrow &\xrightarrow{\text{PT}} L_+^\downarrow \\
L_+^\uparrow &\xrightarrow{\text{T}} L_-^\downarrow
\end{aligned} \tag{2.47}$$

The left-handed superluminal transformations are disjoint in a somewhat different way. By eq. 2.45 the determinants are ±1. However the 0-0 matrix element of eq. 2.7 gives

$$\Lambda_L{}^0{}_0{}^2 - \Sigma_i\, (\Lambda_L{}^i{}_0)^2 = 1 \tag{2.46}$$

The representation of superluminal boosts (eq. 2.14) shows that each factor in eq. 2.46 is imaginary. Thus eq. 2.46 implies

$$\Sigma_i |\Lambda_L{}^i{}_0|^2 \geq 1 \qquad (2.47)$$

$$|\Lambda_L{}^0{}_0| \geq 0 \qquad (2.48)$$

where || indicates absolute value. Thus the magnitude of $\Lambda_L{}^0{}_0$ does not have a gap. Therefore left-handed superluminal transformations can be divided into two categories:

$$\begin{aligned} {}_L L_+: &\quad \det \Lambda_L(\omega, \mathbf{u}) = +1 \\ {}_L L_-: &\quad \det \Lambda_L(\omega, \mathbf{u}) = -1 \end{aligned} \qquad (2.49)$$

Under the PT transformation a left-handed superluminal transformation becomes a right handed superluminal transformation (eq. 2.41b).

Again the various disjoint pieces are related by the discrete transformations of parity P and time reversal T:

$$\begin{aligned} {}_L L_+ &\xrightarrow{P} {}_L L_- \\ {}_L L_+ &\xrightarrow{T} {}_L L_- \end{aligned} \qquad (2.50)$$

2.4 Inhomogeneous Right-Handed Extended Lorentz Group

The inhomogeneous right-handed extended Lorentz group consists of the Lorentz group plus right-handed superluminal transformations that have the form:

$$\Lambda_R(\mathbf{v}, \boldsymbol{\theta}, \mathbf{d}) = \exp[i\omega_R \hat{\mathbf{u}} \cdot \mathbf{K} + i\boldsymbol{\theta} \cdot \mathbf{J} - i\mathbf{d} \cdot \mathbf{P}] \qquad (2.51)$$

in general where $\omega_R = \omega - i\pi/2$.

3. Canonical Spin ½ Tachyon Quantum Field Theory

3.1 Introduction

Tachyons are particles that move faster than the speed of light. As we saw in chapter 0 tachyons exist inside Black Holes, and within current theories – particularly SuperString theories.

Attempts to create canonical tachyon quantum field theories began in the 1960's. These attempts were made within the framework of the Lorentz group and, consequently, were limited to spin 0 theories since there are no finite dimensional representations of the Lorentz group for negative m^2 except for the one-dimensional representation. None of these attempts, or attempts since then, succeeded in creating a canonically quantized spin 0 tachyon quantum field theory.[16]

In this chapter we will formulate a free spin ½ tachyon Quantum Field theory. We choose to develop a spin ½ tachyon theory first because spin ½ particles (quarks and leptons) play an extraordinary role in the Standard Model. In chapter 4 we will consider bosonic tachyons.

We will develop our spin ½ tachyon theory from the "ground up" by applying a left-handed extended Lorentz boost to the Dirac equation, and the Dirac spinor wave function, for a particle at rest. This procedure will give a tachyon spinor wave function, and the momentum space tachyon equation equivalent of the Dirac equation. Then we will obtain the coordinate space tachyon Dirac equation, define a lagrangian, and proceed to create a canonical quantum field theory for spin ½ tachyons.

3.2 Review of a Method of Deriving the Dirac Equation

In this section we will review a method of obtaining the Dirac equation by Lorentz boosts of the spinor wave function of a particle at rest. In the case of a Lorentz

[16] Except Blaha (2006), the first edition of this book.

transformation the 4 x 4 matrix form of a Lorentz transformation of the Dirac matrices is

$$S^{-1}(\Lambda(v))\gamma^{\nu}S(\Lambda(v)) = \Lambda^{\nu}{}_{\mu}(v)\gamma^{\mu} \tag{3.1}$$

where $S(\Lambda(v))$ is

$$S(\Lambda(v)) = \exp(-i\omega\sigma_{0i}v_i/(2|\mathbf{v}|)) = \exp(-\omega\gamma^0\boldsymbol{\gamma}\cdot\mathbf{v}/(2|\mathbf{v}|))$$

1

$$= \cosh(\omega/2)I + \sinh(\omega/2)\gamma^0\boldsymbol{\gamma}\cdot\mathbf{p}/|\mathbf{p}| \tag{3.2}$$

with $\omega = \operatorname{arctanh}(|\mathbf{v}|)$, $\cosh(\omega/2) = [(E+m)/(2m)]^{1/2}$ and $\sinh(\omega/2) = |\mathbf{p}|[2m(E+m)]^{-1/2}$. Also

$$S^{-1}(\Lambda(v)) = \gamma^0 S^{\dagger}(\Lambda(v))\gamma^0 = \exp(\omega\gamma^0\boldsymbol{\gamma}\cdot\mathbf{v}/(2|\mathbf{v}|))$$

$$= \cosh(\omega/2)I - \sinh(\omega/2)\gamma^0\boldsymbol{\gamma}\cdot\mathbf{p}/|\mathbf{p}| \tag{3.3}$$

A generic positive energy plane wave solution of the Dirac equation for a particle at rest with rest energy m is

$$\psi(x) = e^{-imt}w(0) \tag{3.4}$$

with w(0) a four component spinor column vector. It satisfies the momentum space Dirac equation for a particle at rest:

$$(m\gamma^0 - m)e^{-imt}w(0) = 0 \tag{3.5}$$

If we now apply $S(\Lambda(v))$ we find

$$0 = S(\Lambda(v))(m\gamma^0 - m)e^{-imt}w(0) = [mS(\Lambda(v))\gamma^0 S^{-1}(\Lambda(v)) - m]S(\Lambda(v))w(0)$$

A straightforward evaluation shows

$$mS(\Lambda(v))\gamma^0 S^{-1}(\Lambda(v)) = g_{\mu\nu}p^{\mu}\gamma^{\nu} = \not{p} \tag{3.6}$$

where $p^0 = (p^2 + m^2)^{1/2}$, $\mathbf{p} = \gamma m\mathbf{v}$, and $p = |\mathbf{p}|$. In addition

$$S(\Lambda(v))w(0) = w(p) \tag{3.7}$$

is a positive energy Dirac spinor. Therefore the Dirac equation in momentum space has the form:

$$(\not{p} - m)e^{-ip\cdot x}w(p) = 0 \tag{3.8}$$

where the exponential factor, mt, is also boosted to p·x. Eq. 3.8 implies the free coordinate space Dirac equation:

$$(i\gamma^\mu\partial/\partial x^\mu - m)\psi(x) = 0 \tag{3.9}$$

3.3 Derivation of the Tachyonic Dirac Equation

The left-handed superluminal coordinate transformation has the form:

$$\Lambda_L(\omega, \mathbf{u}) = \Lambda(\omega + i\pi/2, \mathbf{u}) = \exp[i\omega_L\hat{\mathbf{u}}\cdot\mathbf{K}] \tag{3.10}$$

where $\omega_L = \omega + i\pi/2$ and

$$\cosh(\omega_L) = i\sinh(\omega) = -\gamma = i\,\gamma_s$$
$$\sinh(\omega_L) = i\cosh(\omega) = -\beta\gamma = i\beta\gamma_s \tag{2.11}$$

with, $\beta = v > 1$, $\gamma_s = (\beta^2 - 1)^{-1/2}$, and $\boldsymbol{\omega \geq 0}$. Thus

$$\sinh(\omega) = \gamma_s$$
$$\cosh(\omega) = \beta\gamma_s \tag{2.12}$$

The corresponding spinor transformation is:

$$S_L(\Lambda_L(\omega, \mathbf{u})) = \exp(-i\omega_L\sigma_{0i}v_i/(2|\mathbf{v}|)) = \exp(-\omega_L\gamma^0\boldsymbol{\gamma}\cdot\mathbf{v}/(2|\mathbf{v}|))$$

$$= \cosh(\omega_L/2)I + \sinh(\omega_L/2)\gamma^0\boldsymbol{\gamma}\cdot\mathbf{p}/|\mathbf{p}| \tag{3.11}$$

The inverse transformation is

$$S_L^{-1}(\Lambda_L(\omega, \mathbf{u})) = \gamma^2\gamma^0 K^{-1}S_L{}^\dagger K\gamma^0\gamma^2 = \gamma^2\gamma^0 S_L{}^T\gamma^0\gamma^2 = \exp(\omega_L\gamma^0\boldsymbol{\gamma}\cdot\mathbf{v}/(2|\mathbf{v}|))$$

$$= \cosh(\omega_L/2)I - \sinh(\omega_L/2)\gamma^0\boldsymbol{\gamma}\cdot\mathbf{p}/|\mathbf{p}| \tag{3.12}$$

where the superscript T denotes the transpose and K is the complex conjugation operator (that also appears in the time-reversal operator). Note that S_L is not unitary just as the equivalent spinor Lorentz transformation $S(\Lambda(v))$ is not unitary.

We can now apply a left-handed superluminal transformation to the generic positive energy plane wave solution of the Dirac equation for a particle of mass m at rest. The result is

$$0 = S_L(\Lambda_L(\omega, \mathbf{u}))(m\gamma^0 - m)e^{-imt}w(0)$$

$$= [mS_L\gamma^0 S_L^{-1} - m]e^{-imt}S_L w(0)$$

where $S_L = S_L(\Lambda_L(\omega, \mathbf{u}))$. After a little algebra

$$mS_L\gamma^0 S_L^{-1} = m[\cosh(\omega_L)\gamma^0 - \sinh(\omega_L)\boldsymbol{\gamma}\cdot\mathbf{p}/|\mathbf{p}|]$$

$$= i\gamma^0 E - i\boldsymbol{\gamma}\cdot\mathbf{p} = i\not{p} \tag{3.13}$$

using eqs. 2.11 and the tachyon energy and momentum expressions

$$\mathbf{p} = m\mathbf{v}\gamma_s \qquad\qquad E = m\gamma_s \tag{3.14}$$

Also

$$S_L w(0) = w_T(p') \tag{3.15}$$

is a tachyon spinor. See Appendix 3-A (at the end of this chapter) for a discussion of tachyon spinors.

The momentum space tachyonic Dirac equation is

$$(i\not{p} - m)e^{ip\cdot x}w_T(p) = 0 \tag{3.16}$$

where $p\cdot x = Et - \mathbf{p}\cdot\mathbf{x}$ after performing a corresponding left-handed superluminal coordinate transformation in the exponential factor based on eq. 2.10d. Thus the positive energy wave is transformed into a negative energy wave by the superluminal transformation.

If we apply i$\not{p}$ to we find the tachyon mass condition is satisfied

$$-E^2 + \mathbf{p}^2 = m^2 \tag{3.17}$$

Transforming back to coordinate space we obtain the *tachyonic Dirac equation*:

$$(\gamma^\mu \partial/\partial x^\mu - m)\psi_T(x) = 0 \tag{3.18}$$

The "missing" factor of i in the first term of eq. 3.18 requires the lagrangian to be different from the conventional Dirac lagrangian in order for the lagrangian to be real. The simplest, physically acceptable, free spin ½ tachyon lagrangian density is:

$$\mathcal{L}_T = \psi_T{}^S(\gamma^\mu \partial/\partial x^\mu - m)\psi_T(x) \tag{3.19}$$

where

$$\psi_T{}^S = \psi_T{}^\dagger i\gamma^0\gamma^5 \tag{3.20}$$

The corresponding action is

$$I = \int d^4x \mathcal{L}_T \tag{3.21}$$

Appendix 3-B proves I is real. The Hamiltonian density is

$$\mathcal{H} = \pi_T \dot{\psi}_T - \mathcal{L} = i\psi_T{}^\dagger\gamma^5(\boldsymbol{\alpha}\cdot\nabla + \beta m)\psi_T = -i\psi_T{}^\dagger\gamma^5\dot{\psi}_T \tag{3.22}$$

using the tachyon Dirac equation to obtain the last equality. The reader will note that the tachyon hamiltonian is hermitean by explicit calculation up to an irrelevant total spatial divergence.

Probability Conservation Law

The tachyon Dirac equation implies a probability conservation law:

$$\partial \rho_5/\partial t = \nabla\cdot\mathbf{j}_5 \tag{3.23}$$

where

$$\rho_5 = \psi_T{}^\dagger\gamma^5\psi_T \qquad \mathbf{j}_5 = \psi_T{}^\dagger\gamma^5\boldsymbol{\alpha}\psi_T \tag{3.24}$$

We are thus led to define the conserved axial charge Q_5

$$Q_5 = \int d^3x\, \psi_T^{\dagger}\gamma^5\psi_T \qquad (3.25)$$

Energy-Momentum Tensor

The tachyon energy-momentum tensor is

$$\mathfrak{T}_{T\mu\nu} = -\, g_{\mu\nu}\,\mathfrak{L}_T + \partial\mathfrak{L}_T/\partial(\partial\psi_T/\partial x_\mu)\, \partial\psi_T/\partial x^\nu \qquad (3.26)$$

$$= i\psi_T^{\dagger}\gamma^0\gamma^5\gamma_\mu\partial\psi_T/\partial x^\nu \qquad (3.27)$$

and thus the conserved energy and momentum are

$$P^0 = H = \int d^3x\, \mathfrak{T}_T^{\;00} = i\int d^3x\psi_T^{\dagger}\gamma^5(\boldsymbol{\alpha}\cdot\nabla + \beta m)\psi_T \qquad (3.28)$$

and

$$P^i = \int d^3x\, \mathfrak{T}_T^{\;0i} = -\, i\int d^3x\, \psi_T^{\dagger}\gamma^5\partial\psi_T/\partial x_i \qquad (3.29)$$

Both the energy and momentum differ significantly from the corresponding quantities for conventional Dirac fields. We leave the calculation of the angular momentum expressions as an exercise for the reader.

3.4 Tachyon Canonical Quantization

Having defined a suitable tachyon lagrangian we can now proceed to its canonical quantization. The conjugate momentum can be calculated from the lagrangian density eq. 3.19:

$$\pi_{Ta} = \partial\, \mathfrak{L}_T/\partial\dot{\psi}_{Ta} \equiv \partial\mathfrak{L}_T/\partial(\partial\psi_{Ta}/\partial t) = -i(\psi_T^{\dagger}\gamma^5)_a \qquad (3.30)$$

The resulting non-zero canonical anti-commutation relations are

$$\{\pi_{Ta}(x), \psi_{Tb}(x')\} = i\,\delta_{ab}\,\delta^3(x - x')$$

or

$$\{\psi_{T\,a}^{\dagger}(x), \psi_{Tb}(x')\} = -\,[\gamma^5]_{ab}\,\delta^3(x - x') \qquad (3.31)$$

At this point we might attempt to complete the canonical quantization procedure in the conventional manner by fourier expanding the field and specifying anti-commutation relations for the fourier component amplitudes. However the incompleteness of the set of plane waves, which are limited by the restriction $|\mathbf{p}| \geq m$, causes the anti-commutator of the fields *not* to yield a $\delta^3(x - x')$. Thus the conventional approach fails to yield the required anti-commutation relations.[17]

Other approaches: 1) decompose the tachyon field into left-handed and right-handed parts and then second quantize each part; and 2) second quantize in light-front coordinates ($x^{\pm} = (x^0 \pm x^3)/\sqrt{2}$). These approaches also both fail.[18]

The only approach that does succeed[19] is to decompose the tachyon field into left-handed and right-handed parts and then second quantize in light-front coordinates. We follow that procedure in the following subsections.

Separation into Left-Handed and Right-Handed Fields

We will use a transformed set of Dirac matrices to develop our left-handed and right-handed tachyon formulations:

$$\gamma^0 = \begin{bmatrix} 0 & -I \\ -I & 0 \end{bmatrix} \qquad \gamma^i = \begin{bmatrix} 0 & \sigma_i \\ -\sigma_i & 0 \end{bmatrix} \qquad \gamma^5 = \begin{bmatrix} I & 0 \\ 0 & -I \end{bmatrix} \qquad (3.32)$$

which are obtained from the usual Dirac matrices by applying the unitary transformation $U = 2^{-\frac{1}{2}}(I + \gamma^5\gamma^0)$. I is the 4×4 identity matrix. The γ^5 chirality operator's eigenvalues define handedness: +1 corresponds to right-handed; and −1 corresponds to left-handed:

$$\gamma^5\psi_L = -\,\psi_L \qquad\qquad \gamma^5\psi_R = \psi_R \qquad (3.33)$$

---

[17] See G. Feinberg, Phys. Rev. **159**, 1089 (1967) for example.

[18] See the first edition Blaha(2006) where these possibilities were considered and found to fail.

[19] Blaha(2006) discusses this case in detail.

Consequently, we can define left-handed and right-handed tachyon fields with the projection operators:

$$\begin{aligned} C^{\pm} &= \tfrac{1}{2}(I \pm \gamma^5) \\ C^{+} + C^{-} &= I \\ C^{\pm\,2} &= C^{\pm} \\ C^{+}C^{-} &= 0 \end{aligned} \qquad (3.34)$$

with the result

$$\begin{aligned} \psi_{TL} &= C^{-}\psi_T \\ \psi_{TR} &= C^{+}\psi_T \end{aligned} \qquad (3.35)$$

We can calculate the commutation relations of the left-handed and right-handed tachyon fields from eq. 3.31 by pre-multiplying and post-multiplying by $\tfrac{1}{2}(1-\gamma^5)$ and $\tfrac{1}{2}(1+\gamma^5)$. The results are:

$$\{\psi_{TLa}^{\dagger}(x), \psi_{TLb}(x')\} = \tfrac{1}{2}(1-\gamma^5)_{ab}\,\delta^3(x-x') \qquad (3.36)$$

$$\{\psi_{TRa}^{\dagger}(x), \psi_{TRb}(x')\} = -\tfrac{1}{2}(1+\gamma^5)_{ab}\,\delta^3(x-x') \qquad (3.37)$$

$$\{\psi_{TLa}^{\dagger}(x), \psi_{TRb}(x')\} = \{\psi_{TRa}^{\dagger}(x), \psi_{TLb}(x')\} = 0 \qquad (3.38)$$

The lagrangian density of eq. 3.19 decomposes into left-handed and right-handed parts:

$$\mathcal{L}_T = \psi_{TL}^{\dagger}\gamma^0 i\gamma^{\mu}\partial_{\mu}\psi_{TL} - \psi_{TR}^{\dagger}\gamma^0 i\gamma^{\mu}\partial_{\mu}\psi_{TR} - im[\psi_{TR}^{\dagger}\gamma^0\psi_{TL} - \psi_{TL}^{\dagger}\gamma^0\psi_{TR}] \qquad (3.39)$$

Further Separation into + and – Light-Front Fields

There have been many studies of light-front (infinite momentum frame) physics in the past forty years.[20] Light-front coordinates cannot be obtained by a

[20] L. Susskind, Phys. Rev. **165**, 1535 (1968); K. Bardakci and M. B. Halpern Phys. Rev. **176**, 1686 (1968), S. Weinberg, Phys. Rev. **150**, 1313 (1966); J. Kogut and D. Soper, Phys. Rev. **D1**, 2901 (1970); J. D. Bjorken, J. Kogut, and D. Soper, Phys. Rev. **D3**, 1382 (1971); R. A. Neville and F. Rohrlich, Nuov. Cim. **A1**, 625 (1971); F. Rohrlich, Acta Phys Austr. Suppl. **8**, 277 (1971); S-J Chang, R. Root, and T-M Yan, Phys. Rev. **D7**, 1133 (1973); S-J Chang, and T-M Yan, Phys. Rev. **D7**, 1147 (1973); T-M Yan, Phys. Rev. **D7**, 1761 (1973); T-M Yan, Phys. Rev. **D7**, 1780 (1973); C. Thorn, Phys. Rev. **D19**, 639 (1979); and references therein.

Lorentz transformation, or by a superluminal transformation, from a standard set of coordinate system variables even in a limiting sense. Instead they are a defined set of variables that have been used to develop quantum field theories that have been shown to be equivalent to quantum field theories based on conventional coordinates. In particular, light-front quantum field theories have been shown to yield fully Lorentz covariant S matrix elements that are the same as S matrix elements calculated in the conventional way.

Light-front variables can be defined by:

$$x^{\pm} = (x^0 \pm x^3)/\surd 2$$

$$\partial/\partial x^{\pm} \equiv \partial^{\mp} \equiv (\partial/\partial x^0 \pm \partial/\partial x^3)/\surd 2 \qquad (3.40)$$

with the "transverse" coordinate variables, x^1 and x^2, unchanged.

The inner product of two 4-vectors has the form

$$x{\cdot}y = x^+y^- + y^+x^- - x^1y^1 - x^2y^2 \qquad (3.41)$$

and the light-front definition of Dirac matrices is:

$$\gamma^{\pm} = (\gamma^0 \pm \gamma^3)/\surd 2 \qquad (3.42)$$

with transverse matrices γ^1 and γ^2 defined as usual. Note the useful identity:

$$\gamma^{\pm\,2} = 0$$

We define "+" and "–" tachyon fields with the projection operators:

$$R^{\pm} = ½(I \pm \gamma^0\gamma^3) \qquad (3.43)$$

Left-handed, ± light-front fields: $\psi_{TL}{}^{\pm} = R^{\pm}C^{-}\psi_T$

Right-handed, ± light-front fields: $\psi_{TR}{}^{\pm} = R^{\pm}C^{+}\psi_T$ (3.44)

Now if we transform to light-front variables and fields as above we obtain the light-front free tachyon lagrangian:

$$\begin{aligned}\mathcal{L}_T = {} & 2^{1/2}\psi_{TL}^{+\dagger} i\partial^- \psi_{TL}^{+} + 2^{1/2}\psi_{TL}^{-\dagger} i\partial^+ \psi_{TL}^{-} - \psi_{TL}^{+\dagger}\gamma^0 i\gamma^j\partial^j\psi_{TL}^{-} - \psi_{TL}^{-\dagger}\gamma^0 i\gamma^j\partial^j\psi_{TL}^{+} - \\ & - 2^{1/2}\psi_{TR}^{+\dagger} i\partial^- \psi_{TR}^{+} - 2^{1/2}\psi_{TR}^{-\dagger} i\partial^+ \psi_{TR}^{-} + \psi_{TR}^{+\dagger}\gamma^0 i\gamma^j\partial^j\psi_{TR}^{-} + \psi_{TR}^{-\dagger}\gamma^0 i\gamma^j\partial^j\psi_{TR}^{+} - \\ & - im[\psi_{TR}^{+\dagger}\gamma^0\psi_{TL}^{-} - \psi_{TL}^{+\dagger}\gamma^0\psi_{TR}^{-} + \psi_{TR}^{-\dagger}\gamma^0\psi_{TL}^{+} - \psi_{TL}^{-\dagger}\gamma^0\psi_{TR}^{+}] \qquad (3.45)\end{aligned}$$

with implied sums over j = 1,2. In contrast to the light-front tachyon lagrangian we note the corresponding light-front Dirac fermion lagrangian is

$$\begin{aligned}\mathcal{L}_D = {} & 2^{1/2}\psi_{L}^{+\dagger} i\partial^- \psi_{L}^{+} + 2^{1/2}\psi_{L}^{-\dagger} i\partial^+ \psi_{L}^{-} - \psi_{L}^{+\dagger}\gamma^0 i\gamma^j\partial^j\psi_{L}^{-} - \psi_{L}^{-\dagger}\gamma^0 i\gamma^j\partial^j\psi_{L}^{+} - \\ & + 2^{1/2}\psi_{R}^{+\dagger} i\partial^- \psi_{R}^{+} + 2^{1/2}\psi_{R}^{-\dagger} i\partial^+ \psi_{R}^{-} - \psi_{R}^{+\dagger}\gamma^0 i\gamma^j\partial^j\psi_{R}^{-} - \psi_{R}^{-\dagger}\gamma^0 i\gamma^j\partial^j\psi_{R}^{+} - \\ & - im[\psi_{R}^{+\dagger}\gamma^0\psi_{L}^{-} + \psi_{L}^{+\dagger}\gamma^0\psi_{R}^{-} + \psi_{R}^{-\dagger}\gamma^0\psi_{L}^{+} + \psi_{L}^{-\dagger}\gamma^0\psi_{R}^{+}] \qquad (3.46)\end{aligned}$$

The difference in signs between eqs. 3.45 and 3.46 will turn out to be a crucial factor in the derivation of features of the Standard Model later.

Returning to the tachyon lagrangian eq. 3.45 we obtain equations of motion through the standard variational techniques:

$$\begin{aligned} 2^{1/2} i\partial^- \psi_{TL}^{+} - \gamma^0 i\gamma^j\partial^j\psi_{TL}^{-} + im\gamma^0\psi_{TR}^{-} &= 0 \\ 2^{1/2} i\partial^- \psi_{TR}^{+} - \gamma^0 i\gamma^j\partial^j\psi_{TR}^{-} + im\gamma^0\psi_{TL}^{-} &= 0 \\ 2^{1/2} i\partial^+ \psi_{TL}^{-} - \gamma^0 i\gamma^j\partial^j\psi_{TL}^{+} + im\gamma^0\psi_{TR}^{+} &= 0 \\ 2^{1/2} i\partial^+ \psi_{TR}^{-} - \gamma^0 i\gamma^j\partial^j\psi_{TR}^{+} + im\gamma^0\psi_{TL}^{+} &= 0 \end{aligned} \qquad (3.47)$$

Eqs. 3.47 show that ψ_{TL}^{-} and ψ_{TR}^{-} are dependent fields that are functions of ψ_{TL}^{+} and ψ_{TR}^{+} on the light-front where x^+ equals a constant. They can be expressed in an integral form as well. (The independent fields ψ_{TL}^{+} and ψ_{TR}^{+} play a fundamental role in tachyon theory and are used to define "in" and "out" tachyon states in perturbation theory.)

The conjugate momenta implied by eq. 3.45 are

$$\pi_{TL}^{+} = \partial\mathcal{L}/\partial(\partial^-\psi_{TL}^{+}) = 2^{1/2} i\psi_{TL}^{+\dagger} \qquad (3.48)$$

$$\pi_{TL}^{\;-} = \partial\mathcal{L}/\partial(\partial^{-}\psi_{TL}^{\;-}) = 0$$
$$\pi_{TR}^{\;+} = \partial\mathcal{L}/\partial(\partial^{-}\psi_{TR}^{\;+}) = -2^{1/2}i\psi_{TR}^{\;+\dagger} \qquad (3.49)$$
$$\pi_{TR}^{\;-} = \partial\mathcal{L}/\partial(\partial^{-}\psi_{TR}^{\;-}) = 0$$

Quantization on surfaces of constant x^+ (light-front surfaces) has been shown to support satisfactory formulations of Quantum Electrodynamics and other quantum field theories. Thus x^+ plays the role of the "time" variable in light-front quantized theories. So we will define canonical equal x^+ anti-commutation relations for spin ½ tachyons.

The resulting canonical equal-light-front ($x^+ = y^+$) anti-commutation relations of the independent fields are:

$$\{\psi_{TL}^{\;+\dagger}{}_a(x), \psi_{TL}^{\;+}{}_b(y)\} = 2^{-1}[C^-R^+]_{ab}\,\delta(x^- - y^-)\delta^2(x - y) \qquad (3.50)$$

$$\{\psi_{TR}^{\;+\dagger}{}_a(x), \psi_{TR}^{\;+}{}_b(y)\} = -2^{-1}[C^+R^+]_{ab}\,\delta(x^- - y^-)\delta^2(x - y) \qquad (3.51)$$

$$\{\psi_{TL}^{\;+\dagger}{}_a(x), \psi_{TR}^{\;+}{}_b(y)\} = \{\psi_{TR}^{\;+\dagger}{}_a(x), \psi_{TL}^{\;+}{}_b(y)\} = 0 \qquad (3.52)$$

$$\{\psi_{TL}^{\;+}{}_a(x), \psi_{TR}^{\;+}{}_b(y)\} = \{\psi_{TR}^{\;+\dagger}{}_a(x), \psi_{TL}^{\;+\dagger}{}_b(y)\} = 0 \qquad (3.53)$$

where the factors of 2^{-1} are the result of the $2^{1/2}$ factor in eqs. 3.48 and 3.49, and the factor of $2^{-1/2}$ in the definition of x^- in eq. 3.40.

If we compare eqs. 3.50 and 3.51 with the corresponding anti-commutation relations of conventional Dirac quantum fields:

$$\{\psi_L^{\;+\dagger}{}_a(x), \psi_L^{\;+}{}_b(y)\} = 2^{-1}[C^-R^+]_{ab}\,\delta(x^- - y^-)\delta^2(x - y) \qquad (3.54)$$

$$\{\psi_R^{\;+\dagger}{}_a(x), \psi_R^{\;+}{}_b(y)\} = 2^{-1}[C^+R^+]_{ab}\,\delta(x^- - y^-)\delta^2(x - y) \qquad (3.55)$$

we see that the right-handed tachyon anti-commutation relation (eq. 3.51) has a minus sign relative to the corresponding right-handed conventional anti-commutation relation (eq. 3.55). The right-handed tachyon anti-commutation relation (eq. 3.51) with its minus sign will require compensating minus signs in its creation and annihilation Fourier component operators' anti-commutation relations.

The sign differences between the lagrangian terms in eqs. 3.47 and 3.48 ultimately lead to parity violating features in the Standard Model lagrangian and thus resolve the long-standing question: Why parity violation? Answer: Nature chooses the Left-handed Eextended Lorentz group. Thus the source of parity violation, and much of the form of the Standard Model, is superluminal physics.

Left-Handed Tachyons

The free, "+" light-front, left-handed tachyon wave function Fourier expansion is:

$$\psi_{TL}^{+}(x) = \sum_{\pm s} \int d^2p dp^+ N_{TL}^{+}(p)\theta(p^+)[b_{TL}^{+}(p, s)u_{TL}^{+}(p, s)e^{-ip\cdot x} + \\ + d_{TL}^{+\dagger}(p, s)v_{TL}^{+}(p, s)e^{+ip\cdot x}] \qquad (3.56)$$

and its hermitean conjugate is

$$\psi_{TL}^{+\dagger}(x) = \sum_{\pm s} \int d^2p dp^+ N_{TL}^{+}(p)\theta(p^+) [b_{TL}^{+\dagger}(p, s)u_{TL}^{+\dagger}(p,s)e^{+ip\cdot x} + \\ + d_{TL}^{+}(p, s)v_{TL}^{+\dagger}(p, s)e^{-ip\cdot x}] \qquad (3.57)$$

where † indicates hermitean conjugate, where

$$N_{TL}^{+}(p) = [m|\mathbf{p}|/((2\pi)^3(p^+(p^+ - p^-) + p_\perp^{\,2}))]^{1/2} \qquad (3.57a)$$

where the anti-commutation relations of the Fourier coefficient operators are

$$\begin{aligned}
\{b_{TL}^{+}(q,s), b_{TL}^{+\dagger}(p,s')\} &= 2^{-1/2}\delta_{ss'}\delta^2(\mathbf{q} - \mathbf{p})\delta(q^+ - p^+) \\
\{d_{TL}^{+}(q,s), d_{TL}^{+\dagger}(p,s')\} &= 2^{-1/2}\delta_{ss'}\delta^2(\mathbf{q} - \mathbf{p})\delta(q^+ - p^+) \\
\{b_{TL}^{+}(q,s), b_{TL}^{+}(p,s')\} &= \{d_{TL}^{+}(q,s), d_{TL}^{+}(p,s')\} = 0 \\
\{b_{TL}^{+\dagger}(q,s), b_{TL}^{+\dagger}(p,s')\} &= \{d_{TL}^{+\dagger}(q,s), d_{TL}^{+\dagger}(p,s')\} = 0 \\
\{b_{TL}^{+}(q,s), d_{TL}^{+\dagger}(p,s')\} &= \{d_{TL}^{+}(q,s), b_{TL}^{+\dagger}(p,s')\} = 0 \\
\{b_{TL}^{+\dagger}(q,s), d_{TL}^{+\dagger}(p,s')\} &= \{d_{TL}^{+}(q,s), b_{TL}^{+}(p,s')\} = 0
\end{aligned} \qquad (3.58)$$

and where the spinors are

$$
\begin{aligned}
u_{TL}^{+}(p, s) &= C^{-} R^{+} S_L(\Lambda_L(\mathbf{p}))w^{1}(0) \\
u_{TL}^{+}(p, -s) &= C^{-} R^{+} S_L(\Lambda_L(\mathbf{p}))w^{2}(0) \\
v_{TL}^{+}(p, s) &= C^{-} R^{+} S_L(\Lambda_L(\mathbf{p}))w^{3}(0) \\
v_{TL}^{+}(p, -s) &= C^{-} R^{+} S_L(\Lambda_L(\mathbf{p}))w^{4}(0) \\
\\
u_{TL}^{+\dagger}(p, s) &= w^{1T}(0)S_L^{\dagger}(\Lambda_L(\mathbf{p}))R^{+}C^{-} \\
u_{TL}^{+\dagger}(p, -s) &= w^{2T}(0)S_L^{\dagger}(\Lambda_L(\mathbf{p}))R^{+}C^{-} \\
v_{TL}^{+\dagger}(p, s) &= w^{3T}(0)S_L^{\dagger}(\Lambda_L(\mathbf{p}))R^{+}C^{-} \\
v_{TL}^{+\dagger}(p, -s) &= w^{4T}(0)S_L^{\dagger}(\Lambda_L(\mathbf{p}))R^{+}C^{-}
\end{aligned}
\tag{3.59}
$$

where the superscript "T" indicates the transpose. (These spinors are described in Appendix 3-A.)

The canonical left-handed, light-front anti-commutation relation (eq. 3.50) follows from eqs. 3.56 – 3.59:

$$
\{\psi_{TL}^{+}{}_{a}(x), \psi_{TL}^{+\dagger}{}_{b}(y)\} = \sum_{\pm s,s'} \int d^2p\,dp^{+} \int d^2p'dp'^{+}\, N_{TL}^{+}(p)N_{TL}^{+}(p')\theta(p^{+})\theta(p'^{+})\cdot
$$

$$
\cdot[\{b_{TL}^{+\dagger}(p',s'),b_{TL}^{+}(p,s)\}u_{TL}^{+}{}_{a}(p,s)u_{TL}^{+\dagger}{}_{b}(p',s')e^{+ip'\cdot y - ip\cdot x} +
$$

$$
+ \{d_{TL}^{+}(p',s'),d_{TL}^{+\dagger}(p,s)\}v_{TL}^{+}{}_{a}(p,s)v_{TL}^{+\dagger}{}_{b}(p',s')e^{-ip'\cdot y + ip\cdot x}]
$$

$$
= \sum_{\pm s} \int d^2p\,dp^{+}\, N_{TL}^{+2}(p)\theta(p^{+})[u_{TL}^{+}{}_{a}(p,s)u_{TL}^{+\dagger}{}_{b}(p,s)e^{+ip\cdot(y-x)} +
$$

$$
+ v_{TL}^{+}{}_{a}(p,s)v_{TL}^{+\dagger}{}_{b}(p,s)e^{-ip\cdot(y-x)}]
$$

$$
\begin{aligned}
= -i\int d^2p\,dp^{+}\theta(p^{+})N_{TL}^{+2}(p)(2m|\mathbf{p}|)^{-1}\{&[C^{-}R^{+}(i\not{p} - m)\gamma\cdot\mathbf{p}R^{+}C^{-}]_{ab}e^{+ip\cdot(y-x)} + \\
&+ [C^{-}R^{+}(i\not{p} + m)\gamma\cdot\mathbf{p}R^{+}C^{-}]_{ab}e^{-ip\cdot(y-x)}\}
\end{aligned}
$$

$$= -i\int d^2p_\perp \int_0^\infty dp^+ N_{TL}^{+2}(p)\{[C^- R^+(ip^+(p^+ - p^-) + ip_\perp{}^2 - mp_\perp\cdot\gamma_\perp)C^-]_{ab}\cdot$$

$$\cdot e^{+ip^+(y^- - x^-) - ip_\perp\cdot(y_\perp - x_\perp)} -$$

$$- [C^- R^+(-ip^+(p^+ - p^-) - ip_\perp{}^2 - mp_\perp\cdot\gamma_\perp)C^-]_{ab} e^{-ip^+(y^- - x^-) + ip_\perp\cdot(y_\perp - x_\perp)}\}/(2m|\mathbf{p}|)$$

$$= \int d^2p_\perp \int_{-\infty}^\infty dp^+ N_{TL}^{+2}(p)[C^- R^+(p^+(p^+ - p^-) + p_\perp{}^2)]_{ab}\cdot$$

$$\cdot e^{+ip^+(y^- - x^-) - ip_\perp\cdot(y_\perp - x_\perp)}/(2m|\mathbf{p}|)$$

upon letting $p^+ \to -p^+$ and $\mathbf{p}_\perp \to -\mathbf{p}_\perp$ in the second term after using $N_{TL}^{+2}(p)(p^+(p^+ - p^-) + p_\perp{}^2) = 1$. The result

$$= \tfrac{1}{2}\int d^2p_\perp \int_{-\infty}^\infty dp^+ (2\pi)^{-3}[C^- R^+]_{ab} e^{+ip^+(y^- - x^-) - ip_\perp\cdot(y_\perp - x_\perp)}$$

$$= 2^{-1}[C^- R^+]_{ab}\,\delta(y^- - x^-)\delta^2(\mathbf{y} - \mathbf{x}) \qquad (3.60)$$

Therefore we have left-handed, light-front quantized tachyons with canonical commutation relations and localized tachyons. As a result we have a canonical tachyon Quantum Field theory.

Right-Handed Tachyons

The case of right-handed tachyons is similar to the left-handed case with only two differences: a minus sign in the creation and annihilation operator anti-commutation relations, and the use of right-handed projection operators. The right-handed tachyon wave function light-front Fourier expansion is:

$$\psi_{TR}^+(x) = \sum_{\pm s} \int d^2p dp^+ N_{TR}^+(p)\theta(p^+)[b_{TR}^+(p, s)u_{TR}^+(p, s)e^{-ip.x} +$$

$$+ \overset{+}{d}_{TR}{}^{\dagger}(p, s)\overset{+}{v}_{TR}(p, s)e^{+ip.x}] \qquad (3.61)$$

and its hermitean conjugate is

$$\overset{+}{\psi}_{TR}{}^{\dagger}(x) = \sum_{\pm s}\int d^2p dp^+ \overset{+}{N}_{TR}(p)\theta(p^+)\,[\overset{+}{b}_{TR}{}^{\dagger}(p, s)\overset{+}{u}_{TR}{}^{\dagger}(p, s)e^{+ip.x} +$$
$$+ \overset{+}{d}_{TR}(p, s)\overset{+}{v}_{TR}{}^{\dagger}(p, s)e^{-ip.x}] \qquad (3.62)$$

where $\overset{+}{N}_{TR}(p) = \overset{+}{N}_{TL}(p)$, where the anti-commutation relations of the Fourier coefficient operators are

$$\{\overset{+}{b}_{TR}(q,s), \overset{+}{b}_{TR}{}^{\dagger}(p,s')\} = -2^{-\frac{1}{2}}\delta_{ss'}\delta^2(\mathbf{q} - \mathbf{p})\delta(q^+ - p^+) \qquad (3.63)$$
$$\{\overset{+}{d}_{TR}(q,s), \overset{+}{d}_{TR}{}^{\dagger}(p,s')\} = -2^{-\frac{1}{2}}\delta_{ss'}\delta^2(\mathbf{q} - \mathbf{p})\delta(q^+ - p^+)$$
$$\{\overset{+}{b}_{TR}(q,s), \overset{+}{b}_{TR}(p,s')\} = \{\overset{+}{d}_{TR}(q,s), \overset{+}{d}_{TR}(p,s')\} = 0$$
$$\{\overset{+}{b}_{TR}{}^{\dagger}(q,s), \overset{+}{b}_{TR}{}^{\dagger}(p,s')\} = \{\overset{+}{d}_{TR}{}^{\dagger}(q,s), \overset{+}{d}_{TR}{}^{\dagger}(p,s')\} = 0$$
$$\{\overset{+}{b}_{TR}(q,s), \overset{+}{d}_{TR}{}^{\dagger}(p,s')\} = \{\overset{+}{d}_{TR}(q,s), \overset{+}{b}_{TR}{}^{\dagger}(p,s')\} = 0$$
$$\{\overset{+}{b}_{TR}{}^{\dagger}(q,s), \overset{+}{d}_{TR}{}^{\dagger}(p,s')\} = \{\overset{+}{d}_{TR}(q,s), \overset{+}{b}_{TR}(p,s')\} = 0$$

and where the spinors are

$$\overset{+}{u}_{TR}(p, s) = \overset{+}{C}\overset{+}{R}u_T(p,s) \qquad (3.64)$$
$$\overset{+}{v}_{TR}(p, s) = \overset{+}{C}\overset{+}{R}v_T(p,s) \qquad (3.65)$$

by Appendix 3-A (eq. 3-A.7).

The right-handed anti-commutation relation (eq. 3.51) with the minus sign follows in particular because of the minus signs in eqs. 3.63.

Interpretation of Tachyon Creation and Annihilation Operators

To properly discuss the physical interpretation of tachyon creation and annihilation operators we must first determine the hamiltonian and momentum operators in terms of creation and annihilation operators.

The energy-momentum tensor density is the symmetrized version of

$$\mathfrak{T}^{\mu\nu} = \sum_i \partial\mathfrak{L}/\partial(\partial\chi_i/\partial x_\mu)\ \partial\chi_i/\partial x_\nu - g^{\mu\nu}\mathfrak{L} \tag{3.66}$$

where the sum over i is over the fields. The light-front hamiltonian is

$$H \equiv P^- = T^{+-} = \int dx^- d^2x \mathfrak{T}^{+-} \tag{3.67}$$

and the "momenta" are

$$P^+ = T^{++} = \int dx^- d^2x \mathfrak{T}^{++} \tag{3.68}$$

$$P^i = T^{+i} = \int dx^- d^2x \mathfrak{T}^{+i} \tag{3.69}$$

for i = 1,2.

The light-front, left-handed and right-handed tachyon lagrangian $\mathfrak{L}_T$ is eq. 3.45 and its equations of motion are eqs. 3.47. They imply

$$H = i2^{-1/2}\int dx^- d^2x\,[\psi_{TL}^{+\dagger}\partial^-\psi_{TL}^{+} - \partial^-\psi_{TL}^{+\dagger}\psi_{TL}^{+} + \psi_{TL}^{-\dagger}\partial^+\psi_{TL}^{-} - \partial^+\psi_{TL}^{-\dagger}\psi_{TL}^{-} -$$
$$- \psi_{TR}^{+\dagger}\partial^-\psi_{TR}^{+} + \partial^-\psi_{TR}^{+\dagger}\psi_{TR}^{+} - \psi_{TR}^{-\dagger}\partial^+\psi_{TR}^{-} + \partial^+\psi_{TR}^{-\dagger}\psi_{TR}^{-} + \text{mass terms}] \tag{3.70}$$

After substituting for the various fields we find the *independent fields* (which constitute the in and out particle states) have the hamiltonian terms:

$$H = \sum_{\pm s}\int d^2p\,dp^+\,p^-[b_{TL}^{+\dagger}(p,s)b_{TL}^{+}(p,s) - d_{TL}^{+}(p,s)d_{TL}^{+\dagger}(p,s) -$$
$$- b_{TR}^{+\dagger}(p,s)b_{TR}^{+}(p,s) + d_{TR}^{+}(p,s)d_{TR}^{+\dagger}(p,s)] \tag{3.71}$$

$$= \sum_{\pm s}\int d^2p\,dp^+\,p^-[b_{TL}^{+\dagger}(p,s)b_{TL}^{+}(p,s) + d_{TL}^{+\dagger}(p,s)d_{TL}^{+}(p,s) -$$
$$- b_{TR}^{+\dagger}(p,s)b_{TR}^{+}(p,s) - d_{TR}^{+\dagger}(p,s)d_{TR}^{+}(p,s)] \tag{3.72}$$

up to the usual infinite constants due to left-handed operator rearrangement and right-handed operator rearrangement that are discarded. Eq. 3.72 is the basis for our particle interpretation of tachyon creation and annihilation operators based on Dirac's hole theory. Dirac hole theory as applied in light-front coordinates assumes all negative p^- ("energy") states are filled.

Left-Handed Tachyon Creation and Annihilation Operators

1. We identify $b_{TL}{}^{+\dagger}(p,s)$ and $d_{TL}{}^{+}(p,s)$ as creation operators for left-handed tachyons. $b_{TL}{}^{+\dagger}(p,s)$ creates a positive p^- ("energy") state and $d_{TL}{}^{+}(p,s)$ creates a negative p^- ("energy") state.

2. $b_{TL}{}^{+}(p,s)$ and $d_{TL}{}^{+\dagger}(p,s)$ are the corresponding annihilation operators for left-handed tachyons. $b_{TL}{}^{+}(p,s)$ annihilates a positive p^- ("energy") state and $d_{TL}{}^{+\dagger}(p,s)$ annihilates a negative p^- ("energy") state.

3. We assume Dirac hole theory holds for the left-handed tachyon vacuum with all negative energy states filled. There is no tachyon energy gap as there is for Dirac fermions. There is also the problem that the left-handed tachyon vacuum is not invariant under ordinary Lorentz transformations or Superluminal transformations. *However if we confine ourselves to light-front coordinates for computations no ambiguity can result and the Lorentz covariant quantities that we calculate, such as the S matrix, are well-defined.*

4. Using tachyon hole theory we identify $b_{TL}{}^{+}(p,s)$ and $d_{TL}{}^{+\dagger}(p,s)$ as annihilation operators for left-handed tachyons. $b_{TL}{}^{+}(p,s)$ annihilates a positive p^- ("energy") state and $d_{TL}{}^{+\dagger}(p,s)$ annihilates a negative p^- ("energy") state – thus creating a hole in the tachyon sea that we view as the creation of a positive p^- ("energy"), left-handed antitachyon. $d_{TL}{}^{+}(p,s)$ annihilates a positive p^- ("energy"), left-handed antitachyon.

Right-Handed Tachyons

The anti-commutation relations of right-handed tachyon creation and annhilation operators (eqs. 3.63) and the right-handed hamiltonian terms have the "wrong" sign compared to corresponding Dirac operators and left-handed tachyon operators. This situation is completely analogous to the situation of time-like photons in the covariant formulation of quantum Electrodynamics.[21] In the case of time-like photons it was possible to introduce an indefinite metric (Gupta-Bleuler formulation), and then to use

---

[21] Bogoliubov (1959) pp. 130-136.

the subsidiary condition $\partial A^{v}/\partial x^{v} = 0$ to reduce the dynamics of QED to the transverse components. Thus the time-like photons were intermediate artifacts needed to have a manifestly covariant formulation while QED observables depended solely on the transverse components of the electromagnetic field.

In the present case of free tachyons, and in leptonic ElectroWeak Theory there is no evident "subsidiary condition" to eliminate the right-handed tachyon fields. But since the only manner in which the right-handed leptonic tachyon fields[22] interact is through mass terms, which can be easily 'integrated out", right-handed leptonic tachyon fields are removed from the observable part of the leptonic ElectroWeak Theory by their "lack of interaction" with left-handed fields.

In the case of quark ElectroWeak Theory right-handed tachyon quark fields have charge (–1/3) and thus experience an electromagnetic interaction as well as a Z interaction. However, since quarks are totally confined, right-handed tachyon quarks will not be able to continuously emit photons or Z's due to energy conservation and their confinement to bound states of fixed positive energy.

Thus right-handed tachyons are analogous to time-like photons in the combined leptonic and quark ElectroWeak Theory – necessary theoretically but prevented from causing a negative energy disaster by the forms of their interactions. We discuss this subject in more detail in the following chapters.

3.3 Tachyon Feynman Propagator

In this section we develop the light-front propagator for tachyons. We begin with a subsection describing the light-front propagators of Dirac fields.

Dirac Field Light-Front Propagators

The light-front Feynman propagator for the ψ^{+} field of a Dirac fermion is

$$iS^{+}{}_{F}(x,y)\gamma^{0} = \theta(x^{+} - y^{+})<0|\psi^{+}(x)\psi^{+\dagger}(y)|0> - \theta(y^{+} - x^{+})<0|\psi^{+\dagger}(y)\psi^{+}(x)|0> \quad (3.73)$$

and does not contain a non-covariant piece due to the projection operators:

$$iS^{+}{}_{F}(x,y) = \int d^{2}p\,dp^{+}\theta(p^{+})[1/(2(2\pi)^{3}p^{+})]\{\theta(x^{+}- y^{+})[R^{+}(\not{p} + m)R^{-}]\, e^{-ip\cdot(x - y)} + $$
$$+ \theta(y^{+}- x^{+})[R^{+}(-\not{p} + m)R^{-}]e^{+ip\cdot(x - y)}\}$$

---

[22] The tachyon fields are provisionally assumed to be neutrino fields in the leptonic sector, and d, s and b quarks in the quark sector.

$$= R^{+} iS_F(x,y)R^{-} \tag{3.74}$$

where $S_F(x,y)$ is the usual Feynman propagator.

The light-front Feynman propagator for a *left-handed* Dirac field ψ^+ is

$$iS^{+}_{LF}(x,y) = \int d^2p dp^+ \theta(p^+)[1/(2(2\pi)^3 p^+)]\{\theta(x^+ - y^+)[C^- R^+(\not p + m)R^- C^-]e^{-ip\cdot(x-y)} + $$
$$+ \theta(y^+ - x^+)[C^- R^+(-\not p + m)R^- C^-]e^{+ip\cdot(x-y)}\}$$

$$= C^- R^{+} iS_F(x,y)R^- C^- \tag{3.75}$$

Tachyon Field Feynman Propagator

Turning now to tachyons, the light-front Feynman propagator for the left-handed $\psi_{TL}^{\ +}$ *tachyon* field is (using the Fourier expansion of the left-handed tachyon field eqs. 3.56 and 3.57):

$$iS^{+}_{TLF}(x,y) = \theta(x^+ - y^+)<0|\psi_{TL}^{\ +}(x)\psi_{TL}^{\ +\dagger}(y)\gamma^0|0> -$$
$$- \theta(y^+ - x^+)<0|\psi_{TL}^{\ +\dagger}(y)\gamma^0\psi_{TL}^{\ +}(x)|0>$$
$$= -i\int d^2p dp^+ \theta(p^+) N_{TL}^{\ +2}(2m|\mathbf{p}|)^{-1} C^- R^+\{\theta(x^+ - y^+)[(i\not p - m)\boldsymbol{\gamma}\cdot\mathbf{p}]e^{-ip\cdot(x-y)} +$$
$$+ \theta(y^+ - x^+)[(i\not p + m)\boldsymbol{\gamma}\cdot\mathbf{p}]e^{+ip\cdot(x-y)}\}R^+C^-\gamma^0$$

If we define the on-shell momentum variable $p_0^{\ -} = (p_0^{\ 1}p_0^{\ 1} + p_0^{\ 2}p_0^{\ 2} - m^2)/(2p_0^{\ +})$, $p_0^{\ +} = p^+$, $p_0^{\ j} = p^j$ (for j = 1, 2), $p_{\perp 0}^{\ 2} = p_0^{\ j}p_0^{\ j}$ and $\not p_0 = p_0\cdot\gamma$ then the above equation can be rewritten as

$$= -iC^- R^+\int d^4p[32\pi^4(p_0^{\ +}(p_0^{\ +} - p_0^{\ -}) + p_{0\perp}^{\ 2})]^{-1}e^{-ip\cdot(x-y)}\ .$$

$$\cdot\{\theta(p^+)(i\not p_0 - m)\boldsymbol{\gamma}\cdot\mathbf{p}_0]/[p^- - p_0^{\ -} + i\varepsilon] +$$

$$+ \theta(-p^+)(i\not p_0 + m)\boldsymbol{\gamma}\cdot\mathbf{p}_0]/[p^- + p_0^{\ -} - i\varepsilon]\}R^+C^-\gamma^0$$

$$= -\tfrac{1}{2}\, i\int d^4p(2\pi)^{-4}[C^- R^+(i\not p - m)\boldsymbol{\gamma}\cdot\mathbf{p}R^+C^-\gamma^0]e^{-ip\cdot(x-y)}.$$

$$\cdot[(p^2 + m^2 + i\varepsilon)(p^+(p^+ - p^-) + p_\perp^{\,2}))]^{-1}$$

using $C^-R^+(i\not{p} - m)\gamma\cdot\mathbf{p}R^+C^- = i\, C^-R^+(p^+(p^+ - p^-) + p_\perp^{\,2})$.

$$= \tfrac{1}{2}C^-R^+\gamma^{+0}\int d^4p(2\pi)^{-4}\, p^+ e^{-ip\cdot(x-y)}/(p^2 + m^2 + i\varepsilon) \qquad (3.76)$$

Similarly the light-front Feynman propagator for the right-handed $\psi_{TR}^{\,+}$ tachyon field is

$$iS^+_{TRF}(x,y) = \theta(x^+ - y^+)\langle 0|\psi_{TR}^{\,+}(x)\psi_{TR}^{\,+\dagger}(y)\gamma^0|0\rangle -$$
$$- \theta(y^+ - x^+)\langle 0|\psi_{TR}^{\,+\dagger}(y)\gamma^0\psi_{TR}^{\,+}(x)|0\rangle$$

$$= -\tfrac{1}{2}C^+R^+\gamma^{+0}\int d^4p(2\pi)^{-4}\, p^+ e^{-ip\cdot(x-y)}/(p^2 + m^2 + i\varepsilon) \qquad (3.77)$$

where the relative minus sign between eqs. 3.76 and 3.77 is due to the relative minus signs of the Fouier component operator anti-commutation relations in eq. 3.58 and 3.63. Thus we find *tachyon* pole terms in the tachyon propagator as one would expect.

Appendix 3-A. Tachyon Spinors

The general form of the solutions of the free tachyon Dirac equation eq. 3.18 can be written

$$\psi_T{}^r(x) = e^{-i\chi_r p\cdot x} w^r(p) \qquad (3\text{-A.1})$$

where $\chi_r = +1$ for r = 1, 2 and $\chi_r = -1$ for r = 3, 4. Denoting the spinors $w^r(p) = w^r(0)$ for a particle is at rest in a frame (E = m) we see they can take the form

$$w^r(0) = \begin{bmatrix} \delta_{1r} \\ \delta_{2r} \\ \delta_{3r} \\ \delta_{4r} \end{bmatrix} \qquad (3\text{-A.2})$$

where Kronecker deltas appear in the brackets. From eq. 3.15 we find

$$S_L(\Lambda_L(\omega, \mathbf{u}))w^r(0) = w_T{}^r(p) \qquad (3\text{-A.3})$$

Using eq. 3.11 for $S_L(\Lambda_L(\omega, \mathbf{u}))$ and

$$\mathbf{p} = m\mathbf{v}\gamma_s \qquad E = m\gamma_s \qquad (3\text{-A.4})$$

we see that eq. 3-A.3 implies the columns of the resulting $S_L(\Lambda_L(\omega, \mathbf{u}))$ matrix are

$$S_L(\Lambda_L(\omega, \mathbf{u})) = \begin{array}{cccc} \underline{w_T^3(p)} & \underline{w_T^4(p)} & \underline{w_T^1(p)} & \underline{w_T^2(p)} \end{array} \\ \begin{bmatrix} \cosh(\omega_L/2) & 0 & \sinh(\omega_L/2)p_z/p & \sinh(\omega_L/2)p_-/p \\ 0 & \cosh(\omega_L/2) & \sinh(\omega_L/2)p_+/p & -\sinh(\omega_L/2)p_z/p \\ \sinh(\omega_L/2)p_z/p & \sinh(\omega_L/2)p_-/p & \cosh(\omega_L/2) & 0 \\ \sinh(\omega_L/2)p_+/p & -\sinh(\omega_L/2)p_z/p & 0 & \cosh(\omega_L/2) \end{bmatrix}$$

(3-A.5)

based on the superluminal transformation of positive energy states to negative energy states (eqs. 3.15 and 3.16) with $p_\pm = p_x \pm ip_y$ and where $p = |\mathbf{p}|$. It is easy to verify

$$(i\not{p} - \chi_r m)w_T^r(p) = 0 \tag{3-A.6}$$

where $\chi_r = -1$ for $r = 1, 2$ and $\chi_r = +1$ for $r = 3, 4$.

The spinors that we defined in eq. 2.10 can be generalized in a manner similar to Dirac spinors. We will use a similar notation to the Dirac spinor notation:

$$\begin{aligned} u_T(p, s) &= w_T^1(p) \\ u_T(p, -s) &= w_T^2(p) \\ v_T(p, s) &= w_T^3(p) \\ v_T(p, -s) &= w_T^4(p) \end{aligned} \tag{3-A.7}$$

We define "double dagger" spinors:

$$\begin{aligned} u_T^{\ddagger}(p, s) &= u_T^{\dagger}(p, s)i\gamma\cdot\mathbf{p}/|\mathbf{p}| \\ u_T^{\ddagger}(p, -s) &= u_T^{\dagger}(p, -s)i\gamma\cdot\mathbf{p}/|\mathbf{p}| \\ v_T^{\ddagger}(p, s) &= v_T^{\dagger}(p, s)i\gamma\cdot\mathbf{p}/|\mathbf{p}| \\ v_T^{\ddagger}(p, -s) &= v_T^{\dagger}(p, -s)i\gamma\cdot\mathbf{p}/|\mathbf{p}| \end{aligned} \tag{3-A.8}$$

.

where † indicates hermitean conjugate, which appear in important spinor "completeness" sums:

$$\sum_{\pm s} u_{T\alpha}(p, s) u_{T}^{\ddagger}{}_{\beta}(p, s) = (2m)^{-1}(i\not{p} - m)_{\alpha\beta} \tag{3-A.9}$$

$$\sum_{\pm s} v_{T\alpha}(p, s) v_{T}^{\ddagger}{}_{\beta}(p, s) = (2m)^{-1}(i\not{p} + m)_{\alpha\beta} \tag{3-A.10}$$

or

$$\sum_{\pm s} u_{T\alpha}(p, s) u_{T}^{\dagger}{}_{\beta}(p, s) = -i(2m)^{-1}[(i\not{p} - m)\gamma\cdot\mathbf{p}/|\mathbf{p}|]_{\alpha\beta} \tag{3-A.11}$$

$$\sum_{\pm s} v_{T\alpha}(p, s) v_{T}^{\dagger}{}_{\beta}(p, s) = -i(2m)^{-1}[(i\not{p} + m)\gamma\cdot\mathbf{p}/|\mathbf{p}|]_{\alpha\beta} \tag{3-A.12}$$

Lastly we define light-front, left-handed tachyon spinors by

$$\begin{aligned}
u_{TL}^{+}(p, s) &= C^{-} R^{+} S_L(\Lambda_L(\omega, \mathbf{u}))w^1(0) \\
u_{TL}^{+}(p, -s) &= C^{-} R^{+} S_L(\Lambda_L(\omega, \mathbf{u}))w^2(0) \\
v_{TL}^{+}(p, s) &= C^{-} R^{+} S_L(\Lambda_L(\omega, \mathbf{u}))w^3(0) \\
v_{TL}^{+}(p, -s) &= C^{-} R^{+} S_L(\Lambda_L(\omega, \mathbf{u}))w^4(0)
\end{aligned} \tag{3-A.13}$$

$$\begin{aligned}
u_{TL}^{+\dagger}(p, s) &= w^{1T}(0)\, S_L^{\dagger}(\Lambda_L(\omega, \mathbf{u}))\, R^{+}C^{-} \\
u_{TL}^{+\dagger}(p, -s) &= w^{2T}(0)\, S_L^{\dagger}(\Lambda_L(\omega, \mathbf{u}))R^{+}C^{-} \\
v_{TL}^{+\dagger}(p, s) &= w^{3T}(0)\, S_L^{\dagger}(\Lambda_L(\omega, \mathbf{u}))R^{+}C^{-} \\
v_{TL}^{+\dagger}(p, -s) &= w^{4T}(0)\, S_L^{\dagger}(\Lambda_L(\omega, \mathbf{u}))R^{+}C^{-}
\end{aligned} \tag{3-A.14}$$

where the superscript "T" indicates the transpose and † indicates hermitean conjugate.

Appendix 3-B. Proof of the Reality of the Tachyon Action

The tachyon lagrangian density and action are

$$\mathcal{L}_T = \psi_T{}^S(\gamma^\mu\partial/\partial x^\mu - m)\psi_T(x) \tag{3.19}$$

$$I = \int d^4x\mathcal{L}_T \tag{3.21}$$

where

$$\psi_T{}^S = \psi_T{}^\dagger\, i\gamma^0\gamma^5 \tag{3.20}$$

The complex conjugate of the tachyon lagrangian density is

$$\mathcal{L}_T{}^* = -\psi_T{}^T\, i\gamma^0\gamma^5(\gamma^{\mu *}\partial/\partial x^\mu - m)\psi_T{}^*(x) \tag{3-B.1}$$

where the superscript ^{T} indicates the transpose. Eq. 3-B.1 can be expressed as a transpose:

$$\mathcal{L}_T{}^* = -i[\psi_T{}^\dagger(\gamma^{\mu\dagger}\overleftarrow{\partial}/\partial x^\mu - m)\gamma^5\gamma^0\psi_T(x)]^T \tag{3-B.2}$$

$$= -i[\psi_T{}^\dagger\gamma^5\gamma^0(-\gamma^\mu\overleftarrow{\partial}/\partial x^\mu - m)\psi_T(x)]^T \tag{3-B.3}$$

$$= [\psi_T{}^\dagger i\gamma^0\gamma^5(-\gamma^\mu\overleftarrow{\partial}/\partial x^\mu - m)\psi_T(x)]^T \tag{3-B.4}$$

$$= \psi_T{}^\dagger i\gamma^0\gamma^5(-\gamma^\mu\overleftarrow{\partial}/\partial x^\mu - m)\psi_T(x) \tag{3-B.5}$$

since eq. 3-B.4 is the transpose of a 1 by 1 matrix. Upon performing a partial integration in the action we find

$$I^* = \int d^4x[\psi_T{}^\dagger i\gamma^0\gamma^5(\gamma^\mu\partial/\partial x^\mu - m)\psi_T(x)] = I \tag{3-B.6}$$

4. Integer Spin Tachyons

4.1 Massive, Scalar Tachyons

The case of massive scalar tachyons would normally be the starting point for the discussion of tachyons. But the importance of spin ½ tachyons in the Standard Model led us to consider spin ½ tachyons first. We now turn to free, neutral, spin 0 tachyons, which we anticipate would satisfy the tachyon equivalent of the Klein-Gordon equation:

$$(\Box - m^2)\phi_T(x) = 0 \tag{4.1}$$

where

$$\Box = \partial/\partial x_\mu \partial/\partial x^\mu \tag{4.2}$$

(The charged scalar tachyon case is analogous.)

Eq. 4.1 can be derived using the canonical procedure from the lagrangian density and action

$$\mathcal{L}_T = \tfrac{1}{2}[\partial\phi_T/\partial x^\mu \partial\phi_T/\partial x_\mu + m^2\phi_T{}^2] \tag{4.3}$$

$$I = \int d^4x \mathcal{L}_T$$

We can canonically second quantize this theory using light-front coordinates. The lagrangian density then becomes

$$\mathcal{L}_T = \partial\phi_T/\partial x^+ \partial\phi_T/\partial x^- - \tfrac{1}{2}\partial\phi_T/\partial x^i \partial\phi_T/\partial x^i + \tfrac{1}{2}m^2\phi_T{}^2 \tag{4.4}$$

The conjugate momentum is

$$\pi_T = \partial\mathcal{L}/\partial(\partial^-\phi_T) = \partial^+\phi_T \equiv \partial\phi_T/\partial x^- \tag{4.5}$$

and the equal x^+ commutation relations[23] are

$$[\pi_T(x), \phi_T(y)] = -i2^{-1/2}\delta(x^- - y^-)\delta^2(\mathbf{x}_\perp - \mathbf{y}_\perp) \tag{4.6}$$

We provisionally define the Fourier expansion of ϕ_T as

$$\phi_T(x) = \int d^2p dp^+ N_T(p)\theta(p^+)[a_T(p)e^{-ip.x} + a_T^\dagger(p)e^{+ip.x}] \tag{4.7}$$

where $N_T(p)$ is

$$N_T(p) = [(2\pi)^3 p^+]^{-1/2} \tag{4.8}$$

and the Fourier component operator *commutation* relations are

$$\begin{aligned} [a_T(q), a_T^\dagger(p)] &= 2^{-1/2}\delta^2(\mathbf{q} - \mathbf{p}')\delta(q^+ - p^+) \\ [a_T(q), a_T(p)] &= [a_T(q), a_T(p)] = 0 \end{aligned} \tag{4.9}$$

We now calculate

$$[\pi_T(x), \phi_T(y)] = [\partial\phi_T(x)/\partial x^-, \phi_T(y)]$$

$$= \int d^2p dp^+ \int d^2p' dp'^+ \, N_T(p)N_T(p')\theta(p^+)\theta(p'^+)\cdot$$

$$\cdot\{-ip^+[a_T(p), a_T^\dagger(p')]e^{+ip'.y - ip.x} + ip^+[a_T^\dagger(p), a_T(p')]e^{-ip'.y + ip.x}\}$$

[23] Feinberg (G. Feinberg, Phys. Rev. **159**, 1089 (1967)) and others have suggested that scalar tachyons obey anti-commutation relations because a Lorentz transformation can change a positive energy to a negative energy (and vice versa). However in light-front coordinates a Lorentz or Superluminal boost in the z direction does not change the sign of the equivalent of energy p^-. Boosts in other directions may change the sign of p^-. However the light-front is a particular choice of variables in a specific frame. Since perturbative and other calculations lead to covariant results we can do all calculations on the light-front, and then, after expressing the results in covariant form, transform to any other reference frame. Then tachyon scattering events seen in the new coordinate system should be in agreement with the corresponding covariant calculation of the event. Therefore scalar tachyon quantization using commutators is acceptable and has the advantage of conforming to the general quantization program for scalar particles.

$$= -i2^{-1/2}\int d^2p_\perp \int_0^\infty dp^+ N_T^{+2}(p)p^+\{e^{+ip^+(y^- - x^-) - ip_\perp\cdot(y_\perp - x_\perp)} + e^{-ip^+(y^- - x^-) + ip_\perp\cdot(y_\perp - x_\perp)}\}$$

$$= -i2^{-1/2}\int d^2p_\perp \int_{-\infty}^\infty dp^+(2\pi)^{-3}\, e^{+ip^+(y^- - x^-) - \mathbf{ip}_\perp\cdot(\mathbf{y}_\perp - \mathbf{x}_\perp)}$$

$$= -i2^{-1/2}\delta(x^- - y^-)\delta^2(\mathbf{x}_\perp - \mathbf{y}_\perp) \tag{4.10}$$

verifying the equal x^+ commutation relation.

Scalar Tachyon Feynman Propagator

The scalar tachyon Feynman propagator is defined by

$$i\Delta_{TF}(x - y) = \theta(x^+ - y^+)\langle 0|\phi_T(x)\,\phi_T(y)|0\rangle + \theta(y^+ - x^+)\langle 0|\phi_T(y)\phi_T(x)|0\rangle$$

$$= \int d^2p dp^+ \int d^2p'dp'^+\, N_T(p)N_T(p')\theta(p^+)\theta(p'^+)\cdot$$

$$\cdot\{\langle 0|a_T(p)a_T^\dagger(p')|0\rangle e^{+ip'.y - ip.x} + \langle 0|a_T(p')a_T^\dagger(p)|0\rangle e^{-ip'.y + ip.x}\}$$

$$= 2^{-1/2}\int d^2p_\perp \int_0^\infty dp^+ N_T^{+2}(p)\{e^{+ip^+(y^- - x^-) - \mathbf{ip}_\perp\cdot(\mathbf{y}_\perp - \mathbf{x}_\perp)} + e^{-ip^+(y^- - x^-) + \mathbf{ip}_\perp\cdot(\mathbf{y}_\perp - \mathbf{x}_\perp)}\}$$

$$= 2^{-1/2}\int d^2p_\perp \int_{-\infty}^\infty dp^+(2\pi)^{-3}\, e^{+ip^+(y^- - x^-) - \mathbf{ip}_\perp\cdot(\mathbf{y}_\perp - \mathbf{x}_\perp)}/p^+$$

$$= -i2^{1/2}\int d^4p(2\pi)^{-4}e^{-ip.(x-y)}/(p^2 + m^2 + i\varepsilon) = i\Delta_{FT}(x - y) \tag{4.11}$$

with the expected tachyon pole term.

4.2 Massive Vector Tachyons

The case of massive vector tachyons is of some interest since massive vector bosons, W and Z bosons, have been found in nature. Therefore there is a possibility that, hitherto undiscovered, massive vector tachyons may exist in nature and might eventually be created by particle accelerators. In this section we will second quantize a massive tachyon vector boson in light-front coordinates.

We begin with a standard, neutral, free, massive vector boson lagrangian with the sign of the mass term changed to make it a tachyon vector boson lagrangian:

$$\mathcal{L}_{\text{TVB}} = -\tfrac{1}{4}\, F_T{}^{\mu\nu}(x) F_{T\mu\nu}(x) - \tfrac{1}{2}\, m^2 V_T{}^{\mu} V_{T\mu} \qquad (4.12)$$

where

$$F_{T\mu\nu} = (\partial V_{T\mu}/\partial x^{\nu} - \partial V_{T\nu}/\partial x^{\mu}) \qquad (4.13)$$

The equations of motion are

$$\partial F_T{}^{\mu\nu}/\partial x^{\nu} - m^2 V_T{}^{\mu} = 0 \qquad (4.14)$$

Eq. 4.14 implies the subsidiary condition

$$\partial V_T{}^{\mu}/\partial x^{\mu} = 0 \qquad (4.15)$$

which, in turn, implies

$$(\Box - m^2) V_T{}^{\mu} = 0 \qquad (4.16)$$

where

$$\Box = \partial/\partial x_{\mu} \partial/\partial x^{\mu} \qquad (4.2)$$

as previously.

Eq. 4.16 is immediately recognizable as a tachyon equation for each component. We now transform the lagrangian to light-front coordinates and proceed to quantize. Using the previous definition of light-front variables we define the fields:

$$F_T{}^{+-} = \partial^+ V_T{}^- - \partial^- V_T{}^+$$
$$F_T{}^{+j} = \partial^+ V_T{}^j - \partial^j V_T{}^+$$

$$
\begin{aligned}
F_T^{-j} &= \partial^- V_T^j - \partial^j V_T^- \\
F_T^{ij} &= \partial^i V_T^j - \partial^j V_T^i \\
V_T^- &= 2^{-1/2}(V_T^0 - V_T^3) \\
V_T^+ &= 2^{-1/2}(V_T^0 + V_T^3)
\end{aligned}
\qquad (4.17)
$$

The light-front equivalent of the lagrangian (eq. 4.12) is:

$$\mathcal{L}_{TVB} = -\tfrac{1}{2}(\partial V_{T\mu}/\partial x^\nu \partial V_T^\mu/\partial x_\nu - \partial V_T^\mu/\partial x_\nu \partial V_{T\nu}/\partial x^\mu) - \tfrac{1}{2}m^2 V_T^\mu V_{T\mu}$$

After using the constraint eq. 4.15 and discarding a total divergence, we see

$$\mathcal{L}_{TVB} \equiv -\tfrac{1}{2}\partial V_{T\mu}/\partial x^\nu \partial V_T^\mu/\partial x_\nu - \tfrac{1}{2}m^2 V_T^\mu V_{T\mu} \qquad (4.18)$$

which becomes

$$
\begin{aligned}
\mathcal{L}_{TVB} \equiv & -\partial^+ V_T^- \partial^- V_T^+ - \partial^+ V_T^+ \partial^- V_T^- + \partial^+ V_T^i \partial^- V_T^i + \partial^i V_T^+ \partial^i V_T^- + \\
& + \tfrac{1}{2}\partial^i V_T^j \partial^i V_T^j - \tfrac{1}{2}m^2(2V_T^+ V_T^- - V_T^i V_T^i)
\end{aligned}
\qquad (4.19)
$$

using light-front coordinates with implied sums over i and j. The resulting equations of motion are

$$
\begin{aligned}
(\Box - m^2)V_T^- &= 0 \\
(\Box - m^2)V_T^+ &= 0 \\
(\Box - m^2)V_T^i &= 0
\end{aligned}
\qquad (4.20)
$$

for i = 1, 2.

The conjugate spacelike-surface momenta are

$$\pi^\mu = \partial\mathcal{L}/\partial(\partial^0 V_T^\mu) = -\,\partial V_T^\mu/\partial x^0 \qquad (4.21)$$

and the conjugate light-front momenta are

$$\pi^{+} = \partial\mathfrak{L}/\partial(\partial^{-}V_{T}^{+}) = -\partial^{+}V_{T}^{-} \equiv -\partial V_{T}^{-}/\partial x^{-} \tag{4.22}$$
$$\pi^{-} = \partial\mathfrak{L}/\partial(\partial^{-}V_{T}^{-}) = -\partial^{+}V_{T}^{+} \equiv -\partial V_{T}^{+}/\partial x^{-} \tag{4.23}$$
$$\pi^{i} = \partial\mathfrak{L}/\partial(\partial^{-}V_{T}^{i}) = \partial^{+}V_{T}^{i} \equiv \partial V_{T}^{i}/\partial x^{-} \tag{4.24}$$

The equal x^{+} commutation relations are

$$[\pi_{T}^{a}(x), V_{T}^{b}(y)] = -i2^{-1/2}\delta^{3ab}(x - y) \tag{4.25}$$

where

$$\delta^{3ab}(x - y) = \int d^{2}k dk^{+} e^{i[k^{+}(x^{-} - y^{-}) - \mathbf{k}.(\mathbf{x} - \mathbf{y})]}[g^{ab} + k^{a}k^{b}/m^{2}]/(2\pi)^{3} \tag{4.26}$$

where $\mathbf{k} = (k^{1}, k^{2})$, and $g^{-+} = g^{+-} = 1 = -g^{11} = = -g^{22}$ with all other $g^{ab} = 0$. The equal x^{+} commutation relations satisfy the constraint:

$$\partial_{a}[\pi_{T}^{a}(x), V_{T}^{b}(y)] = \partial_{b}[\pi_{T}^{a}(x), V_{T}^{b}(y)] = 0 \tag{4.27}$$

implied by eq. 4.15.

Next we define the Fourier expansion of V_{T}^{μ} as

$$V_{T}^{\mu}(x) = \sum_{s}\int d^{2}k dk^{+} N_{TV}(k)\theta(k^{+})\varepsilon^{\mu}(k, s)[a_{T}(k, s)e^{-ik.x} + a_{T}^{\dagger}(k, s)e^{+ik.x}] \tag{4.28}$$

where $k^{2} = 2k^{+}k^{-} - k^{i\,2} = -m^{2}$, and where $N_{TV}(k)$ is

$$N_{TV}(k) = [(2\pi)^{3}k^{+}]^{-1/2} \tag{4.29}$$

There are three spin orientations: two transverse orientations and a longitudinal orientation, $s = \pm 1, 0$. The spin polarization vector satisfies

$$k_{\mu}\varepsilon^{\mu}(k, s) = 0 \tag{4.30}$$

It also satisfies the normalization condition

$$\sum_s \varepsilon^{\mu}(k, s)\varepsilon^{\upsilon}(k, s) = -(g^{\mu\nu} + k^{\mu}k^{\nu}/m^2) \tag{4.31}$$

The Fourier component operator commutation relations are

$$[a_T(q, s), a_T^{\dagger}(p, s')] = 2^{-\frac{1}{2}}\delta_{ss'}\delta^2(\mathbf{q} - \mathbf{p}')\delta(q^+ - p^+) \tag{4.32}$$

$$[a_T(q, s), a_T(p, s')] = [a_T(q, s), a_T(p, s')] = 0$$

Eqs. 4.28 – 4.32 imply the commutation relations eqs. 4.25.

Vector Tachyon Feynman Propagator

The vector tachyon Feynman propagator is defined by

$$i\Delta_{TF}(x - y)^{\mu\nu} = \theta(x^+ - y^+)<0|V_T^{\mu}(x)V_T^{\nu}(y)|0> + \\ + \theta(y^+ - x^+)<0|V_T^{\nu}(y)V_T^{\mu}(x)|0> \tag{4.33}$$

and is equal to

$$= -i\int \frac{d^4k\, e^{-ik\cdot(x-y)}\,(g^{\mu\nu} + k^{\mu}k^{\nu}/m^2)}{(2\pi)^4\,(k^2 + m^2 + i\varepsilon)}$$

The propagator displays the tachyon poles as expected.

4.3 Massive Spin 2 Tachyons – Massive Tachyon Gravitons

Gravitons – the quanta of the gravitation – are massless as far as we know. Massive gravitons have been a subject of a number of theoretical investigations. While there is no evidence for massive gravitons there is evidence that the universe in the large has additional forces at play that affect the rotation of galaxies and seem to be producing an accelerating expansion of the universe. Therefore it is sensible to consider the possibility that massive spin 2 tachyons may exist that could play a role in the understanding of unusual features of the universe in the large. Since the effect of new forces seems to be seen only at distances comparable to the size of galaxies or greater, it is possible that massive spin 2 tachyons may have a small mass of the order of [1/L] where L is the galactic radius of galaxies such as our galaxy.

5. Free Tachyon Discrete Symmetries: C, P, T, and CPT

5.1 Tachyons and the Discrete Symmetries: C, P, and T

The discrete (improper) transformations, parity, time reversal and charge conjugation, are of major importance in analyzing the structure of ElectroWeak Theory and the Standard Model. In this chapter we will examine these transformations with respect to tachyon particles.

First, bosonic tachyons (spins 0, 1, and 2) have discrete transformation properties similar to ordinary bosons and so will not be considered further. The interested reader should read standard texts on this topic and notice the sign of the squared mass does not introduce any distinctive differences between tachyon and normal bosons.

In the case of fermions (odd half integer spin particles) there is a difference between tachyon fermions and conventional fermions. We will consider the case of spin ½ tachyons. Fundamental tachyon fermions of higher spin (should any exist) would also have distinctively different P, C, and T transformation properties.

We will use the manifestly covariant lagrangian

$$\mathcal{L} = \psi_T^{\dagger} i\gamma^0\gamma^5(\gamma^\mu\partial/\partial x^\mu - m)\psi_T(x) \tag{5.1}$$

with equations of motion

$$(\gamma^\mu\partial/\partial x^\mu - m)\psi_T(x) = 0 \tag{5.2}$$

and anti-commutator

$$\{\psi_{T\,a}^{\dagger}(x),\ \psi_{Tb}(x')\} = -\,[\gamma^5]_{ab}\,\delta^3(x - x') \tag{5.3}$$

as the starting points of our discussion.

5.2 Parity

In defining the parity transformation for spin ½ tachyons we try to retain as much similarity as possible to the Dirac spin ½ fermion parity transformation. By definition the parity transformation changes $\mathbf{x} \to -\mathbf{x}$. In the case of a Dirac field if the transformation is defined as:

$$\mathcal{P}\psi(\mathbf{x}, t)\mathcal{P}^{-1} = \gamma^0\psi(-\mathbf{x}, t) \tag{5.4}$$

then the free Dirac field lagrangian, field equation and and anti-commutators are invariant under this transformation.

If we now consider a spin ½ tachyon field and assume the same general form for the transformation:

$$\mathcal{P}\psi_T(\mathbf{x}, t)\mathcal{P}^{-1} = \gamma^0\psi_T(-\mathbf{x}, t) \tag{5.5}$$

then we find

$$\mathcal{P}\mathcal{L}(\mathbf{x}, t)\mathcal{P}^{-1} = -\mathcal{L}(-\mathbf{x}, t) \tag{5.6}$$

$$(\gamma^\mu\partial/\partial x'^\mu - m)\psi_T(-\mathbf{x}, t) = 0 \tag{5.7}$$

$$\{\psi_T{}^\dagger{}_a(x'), \psi_{Tb}(y')\} = [\gamma^5]_{ab}\,\delta^3(x' - y') \tag{5.8}$$

where x' = (−**x**, t) and y' = (−**y**, t). The lagrangian and the anti-commutation relations change sign under the parity transformation. Therefore the physics of tachyons is not invariant under parity. This fact is directly evidenced by eq. 3.45 where the expression of the lagrangian in terms of left-handed and right-handed fields shows the lagrangian changes sign under the interchange of left and right handed fields (an effect of the parity transformation). Thus spin ½ tachyon theory, like nature, violates parity.

Note that parity violation is intrinsic to tachyons – even free tachyons. The discussion of the Standard Model in the following chapters will associate tachyon parity violation with Standard Model parity violation.

5.3 Charge Conjugation

The charge conjugation transformation is connected to the interchange of particle and antiparticle. If we assume that a spin ½ tachyon has charge and is coupled to the electromagnetic field then (assuming the usual gauge coupling) the tachyon lagrangian becomes

$$\mathcal{L} = \psi_T^{\dagger} i\gamma^0\gamma^5[\gamma^\mu(\partial/\partial x^\mu + ieA_\mu) - m]\psi_T(x) \tag{5.9}$$

If the theory were charge conjugation invariant then a unitary operator $\mathcal{C}$ would exist that would change the sign of the electromagnetic current $j^\mu(x)$, and the electromagnetic field A(x, t), while leaving the lagrangian invariant:

$$\mathcal{C} j^\mu(x)\mathcal{C}^{-1} = -j^\mu(x) \qquad ???? \tag{5.10}$$

$$\mathcal{C}\mathbf{A}(\mathbf{x}, t)\mathcal{C}^{-1} = -\mathbf{A}(\mathbf{x}, t) \qquad ???? \tag{5.11}$$

$$\mathcal{C}\mathcal{L}(\mathbf{x}, t)\mathcal{C}^{-1} = \mathcal{L}(\mathbf{x}, t) \qquad ???? \tag{5.12}$$

The tachyon electromagnetic current implied by the lagrangian eq. 5.9 is

$$j_T^\mu(x) = \tfrac{1}{2} e[\psi_T^{\dagger}\gamma^0\gamma^5, \gamma^\mu\psi_T] \tag{5.13}$$

where we antisymmetrize as in the case of Dirac fermions.

We extend the standard charge conjugation transformation[24] *with one modification* from that of a conventional Dirac spin ½ field to the case of spin ½ tachyons:

$$\psi_{TC}(\mathbf{x}, t) = \mathcal{C}_T\psi_T(\mathbf{x}, t)\mathcal{C}_T^{-1} = C_T\bar{\psi}_T^{\,T}(\mathbf{x}, t) = C_T\gamma^0\psi_T^{*}(\mathbf{x}, t) \tag{5.14}$$

where $\psi_{TC}(\mathbf{x}, t)$ is the antitachyon field of opposite charge, where the superscript T indicating the transpose, and where the tachyon charge conjugation matrix (which differs from the Dirac field analogue) is

$$C_T = i\gamma^2\gamma^5\gamma^0 \qquad \text{and} \qquad C_T^{-1} = i\gamma^0\gamma^5\gamma^2 = -C_T = C_T^{\dagger} \tag{5.15}$$

The hermitean conjugate of the antitachyon field is

---

[24] In the Dirac representation. See for example Bjorken (1965) p. 115, or Kaku (1993) p. 117.

$$\psi_{TC}^{\dagger}(\mathbf{x}, t) = \mathcal{C}_T\psi_T^{\dagger}(\mathbf{x}, t)\mathcal{C}_T^{-1} = \bar{\psi}_T^{*}(\mathbf{x}, t)C_T^{-1} = \psi_T^{T}(\mathbf{x}, t)\gamma^0 C_T^{-1} \qquad (5.16)$$

Under this transformation we find the tachyon lagrangian, field equation, and anti-commutator are invariant under charge conjugation:

$$[\gamma^{\mu}(\partial/\partial x^{\mu} + ieA_{\mu}) - m]\psi_{TC}(x) = 0 \qquad (5.17)$$

using $\mathcal{C}_T A_\mu \mathcal{C}_T^{-1} = -A_\mu$, and for the lagrangian in eq. 5.9

$$\mathcal{C}_T\mathcal{L}(\mathbf{x}, t)\mathcal{C}_T^{-1} = \mathcal{L}(\mathbf{x}, t) \qquad (5.18)$$

The charge conjugate anti-commutator is

$$\{\psi_{TC}{}^{\dagger}{}_{a}(x), \psi_{TCb}(y)\} = -[\gamma^5]_{ab}\,\delta^3(x - y) \qquad (5.19)$$

The charge conjugated current

$$j_{TC}{}^{\mu}(x) = \tfrac{1}{2}e[\psi_{TC}{}^{\dagger}\gamma^0\gamma^5, \gamma^{\mu}\psi_{TC}] = -\,j_T{}^{\mu}(x) \qquad (5.20)$$

so that

$$\mathcal{C}_T j_T{}^{\mu}(x)A_\mu\mathcal{C}_T^{-1} = j_T{}^{\mu}(x)A_\mu \qquad (5.21)$$

resulting in the tachyon charge conjugation invariant lagrangian eq. 5.9. Thus tachyons and antitachyons can be expected to have the same charge and mass.

5.4 CP Transformation

The CP transformation has been of major theoretical and experimental interest for some time. Experimentally[25] CP violation has been found in certain sectors: K meson and B meson decays.

Since we have seen that tachyons violate parity and do not violate charge conjugation we can see that tachyons inherently violate CP invariance.

[25] B. Aubert et al, BaBar-PUB-07/001, arXiv:hp-ex/0702046 (2007) and references therein.

5.5 Time Reversal

Time reversal invariance is also a significant theoretical and experimental issue. The standard Dirac fermion time reversal transformation is:

$$\mathfrak{T}\psi(\mathbf{x}, t)\mathfrak{T}^{-1} = T\psi(\mathbf{x}, -t) \tag{5.22}$$

where $\mathfrak{T} = \mathcal{U}K$ where $\mathcal{U}$ is a unitary operator and K is the operator that takes the complex conjugate of all c-numbers, and where

$$T = i\gamma^1\gamma^3 \tag{5.23}$$

Due to the form of $\mathfrak{L}$ (eq. 5.9), which assumes an electromagnetic interaction for the purpose of illustration, we find that the *tachyon* time reversal transformation is

$$\mathfrak{T}_T\psi_T(\mathbf{x}, t)\mathfrak{T}_T^{-1} = T_T\psi_T(\mathbf{x}, -t) \tag{5.24}$$

where $\mathfrak{T}_T = \mathcal{U}_T K$ where $\mathcal{U}_T$ is a unitary operator and K is the operator that takes the complex conjugate of all c-numbers, and where

$$T_T = i\gamma^5\gamma^1\gamma^3 \tag{5.25}$$

The matrix T_T satisfies:

$$T_T^{-1} = -i\gamma^3\gamma^1\gamma^5 = T_T \tag{5.26}$$

$$T_T\gamma^\mu T_T^{-1} = -\gamma_\mu \tag{5.27}$$

Under time reversal the current satisfies

$$\mathfrak{T}_T j_{T\mu}(\mathbf{x}, t)\mathfrak{T}_T^{-1} = j_T^{\ \mu}(\mathbf{x}, -t) \tag{5.28}$$

If we assume the electromagnetic field satisfies

$$\mathfrak{T}\mathbf{A}(\mathbf{x}, t)\mathfrak{T}^{-1} = -\mathbf{A}(\mathbf{x}, -t)$$

under time reversal, then the tachyon lagrangian

$$\mathcal{L} = \psi_T^\dagger i\gamma^0\gamma^5[\gamma^\mu(\partial/\partial x^\mu + ieA_\mu) - m]\psi_T(x) \tag{5.29}$$

satisfies

$$\mathfrak{T}_T\mathcal{L}(\mathbf{x}, t)\mathfrak{T}_T^{-1} = \mathcal{L}(\mathbf{x}, -t) \tag{5.30}$$

under time reversal. Although the action changes by a translation in time, Poincaré translation invariance implies the action is invariant. The equation of motion derived from the lagrangian eq. 5.9 is also invariant under the tachyon time reversal transformation.

Thus we find the dynamics of the tachyon lagrangian theory (eq. 5.9) to be invariant under the tachyon time reversal transformation.

5.6 Tachyon CPT Non-Invariance

The question of CPT invariance has long been of theoretical and experimental interest. For conventional particle theories the CPT Theorem implies CPT invariance under very general conditions. We will examine the case of CPT invariance of a model tachyon theory with lagrangian eq. 5.9. For *Dirac* fermions

$$\mathcal{C}\mathcal{P}\mathfrak{T}\psi_a(\mathbf{x}, t)\mathfrak{T}^{-1}\mathcal{P}^{-1}\mathcal{C}^{-1} = i[\psi^\dagger(-\mathbf{x}, -t)\gamma^5]_a = i[\gamma^5\psi^*(-\mathbf{x}, -t)]_a \tag{5.31}$$

For spin ½ *tachyons*

$$\mathcal{C}_T\mathcal{P}\mathfrak{T}_T\psi_{Ta}(\mathbf{x}, t)\mathfrak{T}_T^{-1}\mathcal{P}^{-1}\mathcal{C}_T^{-1} = -i[\psi^\dagger(-\mathbf{x}, -t)\gamma^5]_a \tag{5.32}$$

where a is a spinor index. More succinctly,

$$\mathcal{C}_T\mathcal{P}\mathfrak{T}_T\psi_T(\mathbf{x}, t)\mathfrak{T}_T^{-1}\mathcal{P}^{-1}\mathcal{C}_T^{-1} = -i\gamma^5\psi^*(-\mathbf{x}, -t) \tag{5.33}$$

Eq. 5.33 differs only by a phase from eq. 5.31.

Therefore one might think that bilinear combinations of ψ and $\psi^\dagger$ which of necessity must be factors in a lagrangian will result in the cancellation of the –1 factors upon CPT transformation.

However the free field lagrangian

$$\mathcal{L} = \psi_T^\dagger(x)i\gamma^0\gamma^5(\gamma^\mu\partial/\partial x^\mu - m)\psi_T(x) \tag{5.34}$$

changes to

$$\mathcal{L}_{CPT} = \psi_T^{\dagger}(-x)i\gamma^0\gamma^5(\gamma^\mu\partial/\partial x^\mu + m)\psi_T(-x) \qquad (5.35)$$

The mass term violates CPT invariance. Thus massive spin ½ tachyons inherently violate CPT invariance. If all spin ½ particles start out massless and acquire masses through spontaneous symmetry breaking then the breaking of CPT invariance by massive, spin ½ tachyons is another consequence of spontaneous symmetry breaking.

5.7 Microcausality and Tachyons

Since the CPT Theorem does not hold for spin ½ tachyons it is of interest to consider Jost's Theorem: CPT invariance is equivalent to weak local commutativity, which is a weak form of microcausality. In the case of spin ½ tachyons *weak local commutativity* (or weak microcausality) is defined as

$$<0|\{\psi_T^{\dagger}(x), \psi_T(y)\}|0> = 0 \quad \text{for} \quad (x-y)^2 < 0$$

(spacelike $(x-y)^2$).

The absence of CPT invariance leads us to inquire if microcausality, and/or weak microcausality, still hold?

To answer this question we evaluate the left-handed field commutator $\{\psi_{TL}^{+\dagger}(x), \psi_{TL}^{+}(y)\}$ to see if the normal microcausality condition holds:

$$\{\psi_{TL}^{+\dagger}(x), \psi_{TL}^{+}(y)\} = 0 \quad \text{for} \quad (x-y)^2 < 0 \qquad (5.36)$$

We insert the Fourier expansions:

$$\{\psi_{TL}{}^{+}{}_a(x), \psi_{TL}{}^{+\dagger}{}_b(y)\} = \sum_{\pm s,s'} \int d^2p dp^+ \int d^2p' dp'^+ N_{TL}{}^{+}(p) N_{TL}{}^{+}(p')\theta(p^+)\theta(p'^+)\cdot$$
$$\cdot[\{b_{TL}{}^{+\dagger}(p',s'), b_{TL}{}^{+}(p,s)\} u_{TL}{}^{+}{}_a(p,s) u_{TL}{}^{+\dagger}{}_b(p',s') e^{+ip'.y - ip.x} +$$
$$+ \{d_{TL}{}^{+}(p',s'), d_{TL}{}^{+\dagger}(p,s)\} v_{TL}{}^{+}{}_a(p,s) v_{TL}{}^{+\dagger}{}_b(p',s') e^{-ip'.y + ip.x}]$$

$$= \sum_{\pm s} \int d^2p dp^+ N_{TL}{}^{+2}(p)\theta(p^+)[u_{TL}{}^{+}{}_a(p,s) u_{TL}{}^{+\dagger}{}_b(p,s) e^{+ip.(y-x)} +$$

$$+ \, v_{TL}{}^{+}{}_{a}(p,s) v_{TL}{}^{+\dagger}{}_{b}(p,s) e^{-ip\cdot(y-x)}]$$

$$= i\int d^2p dp^+ \theta(p^+) N_{TL}{}^{+2}(p)(2m|\mathbf{p}|)^{-1}\{[C^- R^+(-i\not{p} + m)\gamma\cdot\mathbf{p}R^+C^-]_{ab} e^{+ip\cdot(y-x)} + $$
$$+ [C^- R^+(-i\not{p} - m)\gamma\cdot\mathbf{p}R^+C^-]_{ab} e^{-ip\cdot(y-x)}\}$$

$$= \tfrac{1}{2}[C^- R^+]_{ab} \int d^2p_\perp \int_0^\infty dp^+ (2\pi)^{-3} (e^{+ip\cdot(y-x)} + e^{-ip\cdot(y-x)})$$

where $p\cdot(y-x) = p^-(y^+ - x^+) + p^+(y^- - x^-) - \mathbf{p}_\perp\cdot(\mathbf{y}_\perp - \mathbf{x}_\perp)$. Since $p^2 = -m^2$, the integral can be rewritten, after letting $p^\mu = -p^\mu$, as

$$= \tfrac{1}{2}[C^- R^+]_{ab} \int d^2p_\perp \int_{-\infty}^\infty dp^+ (2\pi)^{-3} e^{-ip\cdot(y-x)}$$

where $p^- = (p_\perp{}^2 + m^2)/(2p^+)$. For spacelike $(x - y)^2 < 0$ we can always choose a coordinate system where $y^- - x^- = 0$ with the result

$$\{\psi_{TL}{}^{+\dagger}(x), \psi_{TL}{}^{+}(y)\} = 2^{-1}C^- R^+ \delta(y^- - x^-)\delta^2(\mathbf{y} - \mathbf{x}) \quad \underline{\text{if}} \quad y^- - x^- = 0 \qquad (5.37)$$

Therefore

$$\{\psi_{TL}{}^{+\dagger}(x), \psi_{TL}{}^{+}(y)\} = 0 \quad \text{for} \quad (x - y)^2 < 0 \qquad (5.38)$$

Consequently, free left-handed (or right-handed) tachyons with light-front quantization satisfy the microcausality condition.

Appendix 5-A. Tachyons and the Problem of the Uniformity of the Universe

A major cosmological problem has been the large-scale uniformity of the universe. The standard argument begins with the observation that if we trace back the various regions of the universe to very early times, parts of the universe that are now quite similar could not be in interaction with each other in the distant past because interactions are limited by the speed of light. Thus it is difficult to understand how these regions could be so similar today without any interaction between them that would have established uniformity. One proposed resolution of this difficulty is the inflationary scenario of Guth and others.

However, another solution to the uniformity problem presents itself. If we consider the universe at extremely early times when quarks were not confined, then tachyon quarks and neutrinos could be the mechanism to resolve the uniformity problem since tachyons can exceed the speed of light. A detailed model of this possibility remains to be created.

6. Generalization of the Dirac Equation to Extended Lorentz Group Covariance

6.1 Transformations of Dirac and Tachyon Equations

A Superluminal transformation of the Dirac equation transforms the Dirac equation into the tachyon Dirac equation, and vice versa:

$$S_L(\Lambda_L(\omega, \mathbf{u}))\psi(x) \rightarrow \psi_T'(x') \qquad (6.1a)$$
$$S_L(\Lambda_L(\omega, \mathbf{u}))\psi_T(x) \rightarrow \psi'(x')$$

and

$$S_L(\Lambda_L(\omega, \mathbf{u}))(\gamma^\mu\partial/\partial x^\mu - m)S_L^{-1}(\Lambda_L(\omega, \mathbf{u})) = (i\gamma^\mu\partial/\partial x'^\mu - m) \qquad (6.1b)$$
$$S_L(\Lambda_L(\omega, \mathbf{u}))\gamma^5(i\gamma^\mu\partial/\partial x^\mu - m)\gamma^5 S_L^{-1}(\Lambda_L(\omega, \mathbf{u})) = (\gamma^\mu\partial/\partial x'^\mu - m)$$

where

$$x'^\mu = i\Lambda_L{}^\mu{}_\nu(\omega, \mathbf{u})x^\nu \qquad (6.1c)$$
$$\partial/\partial x'^\mu = -i\Lambda_L{}^\nu{}_\mu(\omega, \mathbf{u})\partial/\partial x^\nu.$$

by eq. 2.10b which in matrix form is

$$X' = E(\mathbf{v})X = i\Lambda_L(\mathbf{v})X \qquad (2.10b)$$

Eqs. 6.1a – 6.1c imply

$$S_L(\Lambda_L(\omega, \mathbf{u}))(\gamma^\mu\partial/\partial x^\mu - m)\psi_T(x) = (i\gamma^\mu\partial/\partial x'^\mu - m)S_L(\Lambda_L(\omega, \mathbf{u}))\psi_T(x)$$
$$= (i\gamma^\mu\partial/\partial x'^\mu - m)\psi'(x') \qquad (6.1d)$$

and

$$S_L(\Lambda_L(\omega, \mathbf{u}))\gamma^5(i\gamma^\mu\partial/\partial x^\mu - m)\psi(x) = (\gamma^\mu\partial/\partial x'^\mu - m)S_L(\Lambda_L(\omega, \mathbf{u}))\gamma^5\psi(x)$$
$$= (\gamma^\mu\partial/\partial x'^\mu - m)\psi_T'(x') \qquad (6.1e)$$

where

$$\psi'(x') = S_L(\Lambda_L(\omega, \mathbf{u}))\psi_T(x) \qquad (6.1f)$$

and

$$\psi_T'(x') = S_L(\Lambda_L(\omega, \mathbf{u}))\gamma^5\psi(x) \qquad (6.1g)$$

Thus the Dirac equation is also not Superluminal covariant.

6.2 Extended Dirac Equation

We will now consider the issue of generalizing the Dirac equation so that the extended equation is covariant under both Lorentz transformations and Superluminal transformations – the set of extended Lorentz transformations.

The only obvious method to obtain an extended Dirac equation that is covariant under extended Lorentz transformations is to define an 8 × 8 matrix generalization. Let

$$đ(x) = \begin{bmatrix} (\gamma^\mu\partial/\partial x^\mu - m) & 0 \\ 0 & (i\gamma^\mu\partial/\partial x^\mu - m) \end{bmatrix} \qquad (6.2)$$

be an 8 × 8 matrix operator with the 4 × 4 matrix elements shown, and let

$$\Psi(x) = \begin{bmatrix} \psi_T(x) \\ \psi(x) \end{bmatrix} \qquad (6.3)$$

be an 8 component column vector composed of a Dirac field and a tachyon field. Then the extended Dirac equation is

$$đ(x)\Psi(x) = 0 \qquad (6.4)$$

We now define the 8 × 8 Superluminal transformation

$$S_{L8}(\Lambda_L(v)) = \begin{bmatrix} 0 & S_L(\Lambda_L(v))\gamma^5 \\ S_L(\Lambda_L(v)) & 0 \end{bmatrix} \qquad (6.5)$$

with inverse transformation

$$S_{L8}^{-1}(\Lambda_L(v)) = \begin{bmatrix} 0 & S_L^{-1}(\Lambda_L(v)) \\ \gamma^5 S_L^{-1}(\Lambda_L(v)) & 0 \end{bmatrix} \tag{6.6}$$

Note: we use the notations $S_L(\Lambda_L(v))$ and $S_L(\Lambda_L(\omega, \mathbf{u}))$ interchaneeably. Applying S_{L8} to eq. 6.4 yields

$$0 = S_{L8}(\Lambda_L(v))đ(x)\Psi(x) = đ(x')\Psi'(x') \tag{6.7}$$

where

$$\Psi'(x') = \begin{bmatrix} S_L\gamma^5\psi(x) \\ S_L\psi_T(x) \end{bmatrix} = \begin{bmatrix} \psi_T'(x') \\ \psi'(x') \end{bmatrix} \tag{6.8}$$

Thus the extended Dirac equation is covariant under generalized Superluminal transformations such as eqs. 6.5 - 6.6. Covariance requires the tachyon and the Dirac particle must have the same absolute value for the mass.

It is easy to show that the extended Dirac equation eq. 6.4 is also covariant under conventional Lorentz transformations in the 8×8 representation:

$$S_8(\Lambda(v)) = \begin{bmatrix} S(\Lambda(v)) & 0 \\ 0 & S(\Lambda(v)) \end{bmatrix} \tag{6.9}$$

with inverse

$$S_8^{-1}(\Lambda(v)) = \begin{bmatrix} S^{-1}(\Lambda(v)) & 0 \\ 0 & S^{-1}(\Lambda(v)) \end{bmatrix} \tag{6.10}$$

and

$$S_{8A}(\Lambda(v)) = \begin{bmatrix} 0 & S(\Lambda(v)) \\ S(\Lambda(v)) & 0 \end{bmatrix} \qquad (6.11)$$

with inverse transformation

$$S_{8A}^{-1}(\Lambda(v)) = \begin{bmatrix} 0 & S^{-1}(\Lambda(v)) \\ S^{-1}(\Lambda(v)) & 0 \end{bmatrix} \qquad (6.12)$$

Under a conventional Lorentz transformation we find

$$0 = S_8(\Lambda(v))\text{đ}(x)\Psi(x) = \text{đ}(x')\Psi'(x') \qquad (6.13)$$

$$0 = S_{8A}(\Lambda(v))\text{đ}(x)\Psi(x) = \text{đ}(x')\Psi'(x')$$

The lagrangian density that corresponds to our 8-dimensional construction is

$$\mathcal{L}_8 = \overline{\Psi}(x)\text{đ}(x)\Psi(x) \qquad (6.14)$$

where

$$\overline{\Psi}(x) = \Psi^\dagger\Gamma^0 \qquad (6.15)$$

and

$$\Gamma^0 = \begin{bmatrix} i\gamma^0\gamma^5 & 0 \\ 0 & \gamma^0 \end{bmatrix} \qquad (6.16)$$

The action is

$$I = \int d^4x\mathcal{L}_8 \qquad (6.17)$$

is invariant under Lorentz transformations S_8 and S_{8A}.

The Hamiltonian density for the 8-dimensional theory is

$$\mathcal{H}_8(x) = \begin{bmatrix} i\psi_T^{\dagger}\gamma^5(\boldsymbol{\alpha}\cdot\nabla + \beta m)\psi_T & 0 \\ 0 & \psi^{\dagger}(-i\boldsymbol{\alpha}\cdot\nabla + \beta m)\psi \end{bmatrix} \tag{6.18}$$

6.3 Non-Invariance of the Extended Free Action Under a Superluminal Transformation

The action 6.17 is not invariant under superluminal transformation. The fundamental cause of this non-invariance is the three dimensional nature of space. In the case of Dirac particles one can define a Lorentz invariant action because time is one-dimensional. Thus one can use $\psi^{\dagger}\gamma^0 = \bar{\psi}$ to form the Dirac field lagrangian and action. A key factor in the Lorentz invariance is the relation between the inverse and hermitean conjugate of the spinor boost operator

$$\gamma^0 S^{-1}\gamma^0 = S^{\dagger} \tag{6.19}$$

In the case of the tachyon lagrangian and action superluminal invariance is not possible because the tachyonic equivalent to eq. 6.19 is[26]

$$S_L^{-1}(\Lambda(\mathbf{v}))\boldsymbol{\gamma}\cdot\mathbf{p}/|\mathbf{p}| = i\gamma^0 S_L^{\dagger}(\Lambda(\mathbf{v})) \tag{6.20}$$

where $\mathbf{p} = m\gamma_s\mathbf{v}$. The appearance of $\boldsymbol{\gamma}\cdot\mathbf{p}/|\mathbf{p}|$ in eq. 6.20 precludes the invariance of the free tachyon action.

We will now show the effect of a superluminal transformation (eqs. 6.5 and 6.6) on the lagrangian density eq. 6.14. The two non-zero parts of the lagrangian density $\mathcal{L}_8$ (eq. 6.14) are

$$\mathcal{L}_1 = \psi_T^{\dagger} i\gamma^0\gamma^5(\gamma^{\mu}\partial/\partial x^{\mu} - m)\psi_T(x) \tag{6.21}$$

and

[26] This relation is derivable from eqs. 3.11 and 3.12.

$$\mathcal{L}_2 = \psi^\dagger\gamma^0(i\gamma^\mu\partial/\partial x^\mu - m)\psi(x) \tag{6.22}$$

The effect of the transformation eqs. 6.5 and 6.6 on these terms is

$$\begin{aligned}\mathcal{L}_1' &= \psi_T^\dagger i\gamma^0\gamma^5 S_L^{-1}S_L(\gamma^\mu\partial/\partial x^\mu - m)\, S_L^{-1}S_L\psi_T(x)\\ &= \psi_T^\dagger i\gamma^0\gamma^5 S_L^{-1}(i\gamma^\mu\partial/\partial x'^\mu - m)S_L\psi_T(x)\\ &= -\psi_T^\dagger S_L^\dagger\gamma^5(\boldsymbol{\gamma}\cdot\mathbf{p}/|\mathbf{p}|)(i\gamma^\mu\partial/\partial x'^\mu - m)S_L\psi_T(x)\\ &= \psi'^\dagger(x')(\boldsymbol{\gamma}\cdot\mathbf{p}/|\mathbf{p}|)\gamma^5(i\gamma^\mu\partial/\partial x'^\mu - m)\psi'(x')\end{aligned} \tag{6.23}$$

using eqs. 6.20, 6.1f and 6.1g; and

$$\begin{aligned}\mathcal{L}_2' &= \psi^\dagger\gamma^0\gamma^5 S_L^{-1}S_L\gamma^5(i\gamma^\mu\partial/\partial x^\mu - m)\gamma^5 S_L^{-1}S_L\gamma^5\psi(x)\\ &= \psi^\dagger\gamma^0\gamma^5 S_L^{-1}(\gamma^\mu\partial/\partial x'^\mu - m)S_L\gamma^5\psi(x)\\ &= i\psi^\dagger\gamma^5 S_L^\dagger(\boldsymbol{\gamma}\cdot\mathbf{p}/|\mathbf{p}|)(\gamma^\mu\partial/\partial x'^\mu - m)S_L\gamma^5\psi(x)\\ &= i\psi_T'^\dagger(x')(\boldsymbol{\gamma}\cdot\mathbf{p}/|\mathbf{p}|)(\gamma^\mu\partial/\partial x'^\mu - m)\psi_T'(x')\end{aligned} \tag{6.24}$$

where $\psi'(x')$ is a solution of the Dirac equation obtained by superluminal boosting (by $\mathbf{p}/m$) of a tachyon field and where $\psi_T'(x')$ is a solution of the Tachyon equation obtained by superluminal boosting (by $\mathbf{p}/m$) of a Dirac field. Eqs. 6.23 and 6.24 clearly show that $\mathcal{L}_8$ is *not* invariant under superluminal transformations.

Consequently the action of eq. 6.17 is only invariant under inhomogeneous Lorentz transformations. *This state of affairs is actually an advantage when we derive features of the Standard Model because it will be seen to prevent any interplay between unbroken ElectroWeak SU(2) rotations and superluminal transformations.*

6.4 The Diracian Dilemma – To What Particles Do They Correspond?

The development of this 8-dimensional formalism, and in particular, the "8-spinor" wave function consisting of a Dirac spinor and and a tachyon spinor, raises the question, "Is there a particle interpretation for the 8-spinor wave function?" Dirac faced a similar issue in 1928-1930 with the negative energy states of the Dirac equation. He developed "hole theory" which eventually led to the interpretation of holes in the sea of filled negative energy as *positrons*. We now face the same problem: with what pairs of particles do we identify the doublets consisting of a Dirac particle and a tachyon?

The obvious natural interpretation of these 8-spinors is ElectroWeak isodoublets such as:

$$\Psi_{L\ell}(x) = \begin{bmatrix} \psi_{\ell TL} \\ \\ \psi_{\ell L} \end{bmatrix} \sim \begin{bmatrix} \nu_L \\ \\ e_L \end{bmatrix} \qquad (6.25)$$

for leptons where "e_L" represents a left-handed charged lepton and ν_L represents a left-handed neutrino. For quarks:

$$\Psi_{Lq}(x) = \begin{bmatrix} \psi_{qL} \\ \\ \psi_{qTL} \end{bmatrix} \sim \begin{bmatrix} u_L \\ \\ d_L \end{bmatrix} \qquad (6.26)$$

where u_L is an up type quark and d_L is a down type quark.[27]

[27] While the lepton situation is clear in the sense that charged leptons cannot be tachyons since their masses are known (Thus only tachyonic neutrinos are the only currently allowed possibility.), the quark situation is somewhat unclear. We have provisionally chosen the "down" type of quark (d, s, and b) as tachyonic. The association of bound states of these quarks such as the K^0 and B^0 systems which are known to have CP violation, and the CP violation engendered by tachyons, encourages this interpretation.

7. Derivation of One Generation Leptonic Standard Model Features

7.1 The Standard Model Leptonic Sector as a Consequence of the Left-handed Extended Lorentz Group

The development of the Standard Model in the 1960's and 1970's was a major step forward in our understanding of the major forces of nature. However the strange, and, in many physicists' opinions, unattractive form of the theory led particle theorists to conclude that it was a provisional theory that would eventually be replaced by a more elegant fundamental theory. Work in these directions has focussed upon 1) embedding the Standard Model within a larger ("elegant") symmetry group; 2) embedding the Standard Model within a space-time with extra dimensions that generate the Standard Model through some mechanism such as the Kaluza-Klein mechanism; and 3) viewing the Standard Model as somehow a "low energy" phenomenology that emerges from a Superstring Theory.

In this chapter we will take an alternate view. We will derive the leptonic sector of the Standard Model based on Left-handed Extended Lorentz group covariance of the equations of motion. Thus, unlike previous efforts, we view the Standard Model as in a natural form dictated by Left-handed Extended Lorentz group covariance and certain other fundamental physical requirements. Beauty being in the eye of the beholder, we shall endeavor to show the attractiveness of the derivation of the Standard Model from covariance based on the extension of the Lorentz group beyond the speed of light.

The naturalness of the derivation, and its close connection to the left-handed superluminal extension of the Lorentz group, strongly suggest the Standard Model, which was grown by theorists from experiment, has an undeniable quality of genuiness that will likely survive the passage of time. The basis of the derivation in a more fundamental theoretical framework raises the hope that we have found a newer, deeper level of understanding of elementary particles.

7.2 Assumptions for the Leptonic Sector of the Standard Model

We will make certain assumptions that provide a basis for the derivation of most aspects of the leptonic sector the Standard Model (The quark sector is derived in a subsequent chapter.):

1. The equations of motion of the unbroken form of the Standard Model are largely determined by covariance under the Left-handed Extended Lorentz group, and local gauge symmetries. Electron number is not gauged.
2. Leptonic matter is composed of spin ½ Dirac particles with charge -1 and tachyons of charge zero as well as their anti-particles.[28]
3. Neutrinos are tachyons with a non-zero bare mass that is not altered by the Higgs mechanism but may be renormalized to their physical mass values.
4. Gauge fields are massless "before" spontaneous symmetry breaking and are thus conventional gauge fields without a tachyon equivalent in the theory.
5. Left-handed Extended Lorentz group covariance and gauge symmetries are spontaneously broken through the appearance of mass terms generated by a mechanism such as the Higgs mechanism.
6. One generation of leptons is assumed.

7.3 Derivation of the Leptonic Sector of the Standard Model

The steps of the derivation are:

A. Left-handed Extended Lorentz Group covariance of the dynamical equations of motion requires that spin ½ particles be described by a generalization of the Dirac equation to an 8×8 matrix form in eqs 6.2 – 6.18 based on a doublet consisting of a Dirac particle and a tachyon.

B. We identify the Dirac particle with a charged lepton and the tachyon with a neutrino. The bare masses of these particles has the same numeric value (before symmetry breaking).

C. The leptonic sector free field lagrangian is explicitly

[28] If neutrinos are Majorana particles then the derivation must be modified.

$$\mathcal{L}_{\text{freelep}} = \Psi^\dagger(x)\begin{bmatrix} \gamma^0\gamma^5 i(\gamma^\mu\partial/\partial x^\mu - m) & 0 \\ 0 & \gamma^0(i\gamma^\mu\partial/\partial x^\mu - m)\end{bmatrix}\Psi(x) \qquad (7.1)$$

Focussing on the derivative term we see that it can be put in the form

$$\Psi^\dagger(x)\gamma^0\begin{bmatrix}\gamma^5 & 0\\ 0 & I_4\end{bmatrix} i\gamma^\mu\partial/\partial x^\mu\,\Psi(x) = \Psi^\dagger(x)\gamma^0[C^+I - C^-\sigma_3]i\gamma^\mu\partial/\partial x^\mu\Psi(x) \qquad (7.2)$$

where I_4 is a 4×4 identity matrix, where I and σ_3 are 2 × 2 matrices, and where C^+ and C^- are defined in eq. 3.34. The Pauli matrices σ_i are

$$\sigma_1 = \begin{bmatrix} 0 & 1\\ 1 & 0\end{bmatrix} \qquad \sigma_2 = \begin{bmatrix} 0 & -i\\ -i & 0\end{bmatrix} \qquad \sigma_3 = \begin{bmatrix} 1 & 0\\ 0 & -1\end{bmatrix} \qquad (7.3)$$

Expression 7.2 can be re-expressed in terms of left-handed and right-handed fields as

$$\Psi_L^\dagger(x)\gamma^0 i\gamma^\mu\partial/\partial x^\mu\Psi_L(x) - \Psi_R^\dagger(x)\gamma^0 i\gamma^\mu\partial/\partial x^\mu\sigma_3\Psi_R(x) \qquad (7.4)$$

D. At this point we are in a position to introduce couplings to gauge fields. In view of the doublet nature of the fields $\Psi_L(x)$ and $\Psi_R(x)$ it would appear, at first glance, that the symmetry group of the gauge fields would be $SU(2)_L \otimes SU(2)_R$. However the right-handed tachyon field in expression 7.4 has the wrong sign in the lagrangian, as has been noted in the previous discussion of the free tachyon lagrangian and anti-commutator. Consequently the right-handed tachyon field *cannot* have trilinear or higher order couplings. If it did have interactions then it would rapidly degrade to lower and lower energy by the emission of particles since right-handed leptonic tachyons can exist in principle as free particles (modulo possible Higgs terms). (In this regard the situation is similar to that of time-like photons, except that the set of tachyon physical states cannot be defined in a manner analogous to Gupta-Bleuler electrodynamics where the timelike and longitudinal photons "cancel" each other so that only transverse photons have physical effects.) *Thus there must be no right-handed leptonic tachyon interactions*[29] *(thus excluding the Higgs mechanism for neutrino massess but not mass renormalization.)*

The doublet nature of the left-handed sector implies at least a local SU(2) symmetry implemented with a covariant derivative.

The restricted nature of the right-handed leptonic sector indicates that only the Dirac particle in the "right-handed doublet" can have an interaction. Also the appearance of σ_3 in the right-handed term in expression 7.4 breaks SU(2) invariance <u>if</u> the left-handed covariant derivative (eq. 7.5 below) were substituted for $\partial/\partial x^\mu$ in the right-handed term. Thus a U(1) local gauge field interaction, restricted to the Dirac field member of the right-handed doublet, is the only allowed possibility. Without it, the "right-handed doublet" would have no trilinear or higher order interactions and would be physically irrelevant in the unbroken gauge theory before spontaneous breakdown.

Putting these symmetries together we obtain a left-handed covariant derivative implementing local SU(2)⊗U(1) invariance:

$$D_{L\mu} = \partial/\partial x^\mu + \tfrac{1}{2} i g_2 \boldsymbol{\sigma}\cdot\mathbf{W}_\mu + \tfrac{1}{2} i g' B_\mu \qquad (7.5)$$

and a right-handed covariant derivative[30]

$$D_{R\mu} = \partial/\partial x^\mu \sigma_3 + \tfrac{1}{2} i g' B_\mu |Q| \sigma_3$$

$$= \partial/\partial x^\mu \sigma_3 + \tfrac{1}{2} i g' B_\mu |Q| \qquad (7.6)$$

where Q is the charge operator using $|Q|\sigma_3 = |Q|$ for leptons. We use the absolute value of Q in order to achieve consistency in form with the right-handed quark sector described in the next chapter. As a result expression 7.4 becomes

---

[29] Right-handed neutrinos must interact with gravitons due to their mass-energy and the universality of the gravitational interaction. The weakness of the gravitational interaction mitigates this effect.

[30] The coupling constants are defined by $e = -g'\cos\theta_W = g_2\sin\theta_W$.

$$\Psi_L^{\dagger}(x)\gamma^0 i\gamma^{\mu} D_{L\mu}\Psi_L(x) - \Psi_R^{\dagger}(x)\gamma^0 i\gamma^{\mu} D_{R\mu}\Psi_R(x) \qquad (7.7)$$

Thus the leptonic sector of the lagrangian[31] (modulo mass/Higgs terms) is

$$\mathcal{L}_{lep1} = \Psi_L^{\dagger}\gamma^0 i\gamma^{\mu} D_{L\mu}\Psi_L - \Psi_R^{\dagger}\gamma^0 i\gamma^{\mu} D_{R\mu}\Psi_R \qquad (7.8)$$

$$= \Psi_L^{\dagger}\gamma^0 i\gamma^{\mu} D_{L\mu}\Psi_L + \overline{\psi}_{eR}\gamma^0 i\gamma^{\mu} D_{R\mu}\psi_{eR} - \overline{\psi}_{\upsilon R}\gamma^0 i\gamma^{\mu}\partial/\partial x^{\mu}\psi_{\upsilon R}$$

$$= \Psi_L^{\dagger}\gamma^0 i\gamma^{\mu} D_{L\mu}\Psi_L + \overline{\psi}_{eR}\gamma^0 i\gamma^{\mu}(\partial/\partial x^{\mu} + \tfrac{1}{2}ig'B_{\mu})\psi_{eR} - \overline{\psi}_{\upsilon R}\gamma^0 i\gamma^{\mu}\partial/\partial x^{\mu}\psi_{\upsilon R}$$

where we identify the tachyon as a neutrino and the Dirac particle as a charged lepton such as the electron. Our leptonic sector lagrangian is now the usual leptonic sector ElectroWeak lagrangian with a tachyonic neutrino.

E. Gauge invariance prior to symmetry breaking: The gauge field sector has the usual Yang-Mills lagrangian terms and the B field has lagrangian terms similar to that of the QED lagrangian.

F. Spontaneous symmetry breaking of gauge symmetry, and of Extended Lorentz group covariance, via the Higgs mechanism can be implemented in such a way as to give the electron its known mass as well as the massive vector bosons. Since spontaneous symmetry breaking breaks Left-handed Extended Lorentz covariance to Lorentz covariance it is a moot point whether the Higgs sector exhibits a similar covariance.

G. The origin of the three generations of leptons is not resolved by this derivation.

That concludes the derivation of leptonic sector of the Standard Model. We have shown that the form of the leptonic sector of the Standard Model is fundamental in nature and based on Left-handed Extended Lorentz group covariance of the equations of motion. Thus it may not be correct to view the Standard Model as the result of the breakdown of a larger internal symmetry group.

---

[31] Note that the gauge fields do not appear with a tachyon equivalent since they are initially massless prior to spontaneous symmetry breaking.

8. Can Neutrinos Be Tachyons?

Recently the three species of neutrinos have been found to have masses in neutrino oscillation experiments although the masses are very small. If we denote the three neutrinos as m_1, m_2, and m_3; and let $\Delta m_{ij}^2 = m_j^2 - m_i^2$, then[32]

$$\Delta m_{12}^2 \cong 8 \times 10^{-5}\ \mathrm{eV}^2$$

$$\Delta m_{23}^2 \cong 2.8 \times 10^{-3}\ \mathrm{eV}^2$$

It is claimed that the sign of each Δm^2 is unambiguously determined.[33] However the mass difference values so obtained can be interpreted as differences in tachyon masses as well as differences in "normal" particle masses. The dependence of neutrino oscillations on masses squared results from the time evolution of mixed neutrino states. We consider mixtures of two neutrino states in the vacuum (for the sake of simplicity):

$$|\nu_a\rangle = \cos\theta\, |\nu_e\rangle - \sin\theta\, |\nu_\mu\rangle$$

$$|\nu_b\rangle = \sin\theta\, |\nu_e\rangle + \cos\theta\, |\nu_\mu\rangle$$

with differing masses m_a and m_b.[34] The phase factors determining the time dependence of $|\nu_a\rangle$ and $|\nu_b\rangle$ generate the neutrino oscillations from which the mass relations were obtained. The phase of the kth state is

$$|\nu_k(t)\rangle \sim e^{-im_k^2 t/(2p)}$$

for normal neutrinos. In the case of tachyons the factor in the exponential changes sign:

---

[32] S. M. Bilenky, "Neutrino Masses, Mixing and Oscillations", arXiv:hep-ph/050175 (October 13, 2005).
[33] A. B. McDonald, "Evidence for Neutrino Oscillations I", p. 8 arXiv:nucl-ex/0412005 (December, 2005).
[34] L. Wolfenstein, Phys. Rev. **D17**, 2369 (1978).

Tachyonic Neutrinos

$$E = (\mathbf{p}^2 - m^2)^{1/2} \cong p - m^2/(2p)$$

"Normal" Neutrinos

$$E = (\mathbf{p}^2 + m^2)^{1/2} \cong p + m^2/(2p)$$

Thus tachyonic neutrinos would exhibit the time dependence

$$|\nu_k(t)> \sim e^{+im_{Tk}{}^2 t/(2p)}$$

where we use the mass subscript "T" to indicate a tachyon.

As a result, if neutrinos are "normal" the experimental results above would suggest the neutrino masses satisfy

$$m_3{}^2 > m_2{}^2 > m_1{}^2$$

On the other hand, if neutrinos are tachyons the experimental results above would suggest the value of the tachyon neutrino masses satisfies

$$m_1{}^2 > m_2{}^2 > m_3{}^2$$

an inverted spectrum compared to "normal" neutrinos. When one considers the fact that tachyon masses squared are negative we see that an ordering of the tachyon neutrino mass spectrum, consistent with experimental neutrino results, is:

A Negative Neutrino m^2 Spectrum

$m^2 = 0$ ---------

$m_1{}^2$----------

$m_2{}^2$----------

$m_3{}^2$-----------

9. Derivation of One Generation Quark Sector Standard Model Features

9.1 Quark Sector of the Standard Model

The derivation of the form of the quark sector of the Standard Model is very similar to the preceding derivation of the form of the leptonic sector – but with some important points of difference.

9.2 Quark Sector Assumptions

We will make assumptions that will provide a basis for a derivation of most aspects of single generation, quark sector Standard Model:

1. The equations of motion of the unbroken form of the Standard Model are largely determined by covariance under the Left-handed Extended Lorentz group, and local gauge symmetries. Baryon number is not gauged.
2. Quarks are composed of spin ½ Dirac particles and tachyons.
3. Spin ½ asymptotic baryon states must exist.
4. Gauge fields are massless "before" spontaneous symmetry breaking and are thus conventional gauge fields without a tachyon equivalent in the theory.
5. Left-handed Extended Lorentz group covariance and gauge symmetries are spontaneously broken through the appearance of mass terms generated by a mechanism such as the Higgs mechanism.
6. One generation of quarks is assumed.

9.3 Derivation of the Form of the Standard Model Quark Sector

The derivation:

A. Left-handed Extended Lorentz Group covariance of the dynamical equations of motion requires that spin ½ particles be described by a generalization of the Dirac equation to an 8 × 8 matrix form similar to eqs 6.2 – 6.18 based on a doublet consisting of a Dirac particle and a tachyon. However in the case of quarks the Dirac particle is the top component in the doublet.

B. Thus the 8 × 8 quark matrix formalism is:

$$đ_q(x) = \begin{bmatrix} (i\gamma^\mu\partial/\partial x^\mu - m_0) & 0 \\ 0 & (\gamma^\mu\partial/\partial x^\mu - m_0) \end{bmatrix} \tag{9.1}$$

$$đ_q(x)\Psi_q(x) = 0 \tag{9.2}$$

where

$$\Psi_q(x) = \begin{bmatrix} \psi(x) \\ \psi_T(x) \end{bmatrix} \tag{9.3}$$

The upper 4-component field is a Dirac quark field and the lower 4-component field is a tachyon quark field. The generalized fermion equation eq. 9.2 is covariant under 8 × 8 doublet Superluminal transformations:

$$S_{L8q}(\Lambda_L(v)) = \begin{bmatrix} 0 & S_L(\Lambda_L(v)) \\ S_L(\Lambda_L(v))\gamma^5 & 0 \end{bmatrix} \tag{9.4}$$

with inverse transformation

$$S_{L8q}{}^{-1}(\Lambda_L(v)) = \begin{bmatrix} 0 & \gamma^5 S_L{}^{-1}(\Lambda_L(v)) \\ S_L{}^{-1}(\Lambda_L(v)) & 0 \end{bmatrix} \tag{9.5}$$

Note: we use the notations $S_L(\Lambda_L(v))$ and $S_L(\Lambda_L(\omega, \mathbf{u}))$ interchangeably. Applying S_{L8} to eq. 9.2 yields

$$0 = S_{L8q}(\Lambda_L(v))đ_q(x)\Psi_q(x) = đ_q(x')\Psi_q{}'(x') \qquad (9.6)$$

where

$$\Psi'(x') = \begin{bmatrix} S_L\psi_T(x) \\ S_L\gamma^5\psi(x) \end{bmatrix} = \begin{bmatrix} \psi'(x') \\ \psi_T{}'(x') \end{bmatrix} \qquad (9.7)$$

The generalized fermion equation eq. 9.2 is also covariant under conventional Lorentz transformations in the 8×8 representation.

The free quark sector lagrangian density that corresponds to the 8-dimensional fermion equation eq. 9.2 is

$$\mathscr{L}_{freeQuark} = \overline{\Psi}_q(x)đ_q(x)\Psi_q(x) \qquad (9.8)$$

where

$$\overline{\Psi}_q(x) = \Psi_q{}^\dagger\Gamma_q{}^0 \qquad (9.9)$$

and

$$\Gamma_q{}^0 = \begin{bmatrix} \gamma^0 & 0 \\ 0 & i\gamma^0\gamma^5 \end{bmatrix} \qquad (9.10)$$

C. Having seen that Superluminal covariance requires doublets composed of a tachyon and a Dirac particle with the same absolute value for the bare mass m_0 we now **identify the Dirac particle with a "u-type" quark and the tachyon with a "d-type" quark (an arbitrary choice[35])**. If we assume one quark generation, and do not introduce gauge fields as yet, then we have the free field lagrangian explicitly as

[35] Since CP violation seems to be associates with d, s, and b bound states such as the K^0 and B^0 (and their antiparticle) systems, and since tachyon dynamics violates CP invariance there is some justification for this choice although a detailed analysis remains to be done.

$$\mathcal{L}_{\text{freeQuark}} = \Psi_q^{\dagger}(x)\begin{bmatrix} \gamma^0(i\gamma^{\mu}\partial/\partial x^{\mu} - m_0) & 0 \\ 0 & \gamma^0\gamma^5 i(\gamma^{\mu}\partial/\partial x^{\mu} - m_0) \end{bmatrix}\Psi_q(x) \qquad (9.11)$$

with

$$\Psi_q(x) = \begin{bmatrix} \psi_u \\ \psi_{Td} \end{bmatrix} \qquad (9.12)$$

Focussing on the derivative term we see that it can be put in the form

$$\Psi_q^{\dagger}(x)\gamma^0\begin{bmatrix} I_4 & 0 \\ 0 & \gamma^5 \end{bmatrix} i\gamma^{\mu}\partial/\partial x^{\mu}\,\Psi_q(x) = \Psi_q^{\dagger}(x)\gamma^0[C^+I + C^-\sigma_3]i\gamma^{\mu}\partial/\partial x^{\mu}\Psi_q(x) \qquad (9.13)$$

where I_4 is a 4×4 identity matrix, where I, and the Pauli matrix σ_3, are 2 × 2 matrices, and C^+ and C^- are defined in eq. 3.34. The Pauli matrices σ_i are defined by eq. 7.3.

Expression 9.13 can be expressed in terms of left-handed and right-handed fields as

$$\Psi_{qL}^{\dagger}(x)\gamma^0 i\gamma^{\mu}\partial/\partial x^{\mu}\Psi_{qL}(x) + \Psi_{qR}^{\dagger}(x)\gamma^0 i\gamma^{\mu}\partial/\partial x^{\mu}\sigma_3\Psi_{qR}(x) \qquad (9.14)$$

D. At this point we are in a position to introduce couplings to gauge fields. In view of the doublet nature of the fields $\Psi_{qL}(x)$ and $\Psi_{qR}(x)$ it would again appear, at first glance, that the symmetry group of the gauge fields would be $SU(2)_L \otimes SU(2)_R$. However the right-handed tachyon field in expression 9.14 has the wrong sign in the lagrangian, as has been noted in the previous discussion of the free tachyon lagrangian and anti-commutator. Consequently, **if quarks were *not* confined,** the right-handed quark tachyon field could not have trilinear or higher order couplings. **If free tachyon quarks existed, and had interactions, then they would rapidly degrade to lower and lower energy by the emission of particles.**

However, because of quark confinement in bound states with discrete energy levels, a bound tachyon quark can only emit particles if a lower energy bound state exists. As a result right-handed tachyon quarks can have interactions, such as the electromagnetic interaction, because quark confinement "tames" their propensity to emit particles due to the "wrong sign" in the lagrangian.

Again there is an analogy to Gupta-Bleuler QED quantization. In Gupta-Bleuler quantization physical states are required to have equal numbers of time-like and longitudinal photons thus canceling their physical effects. Similarly, right-handed tachyon quarks are required to be bound to other quarks by quark confinement to avoid continuous emission of particles.[36] Since interactions are allowed for right-handed tachyon quarks the Higgs mechanism can be used to change their mass.

The doublet nature of the left-handed sector implies at least a local SU(2) symmetry. The appearance of σ_3 in the right-handed term in expression 9.14 breaks SU(2) invariance if the left-handed covariant derivative (eq. 9.15 below) were substituted for $\partial/\partial x^\mu$ in the right-handed term. Thus the right-handed fields can only have a U(1) local gauge field interaction, and are SU(2) singlets. Together we thus obtain a left-handed covariant derivative implementing local SU(2)⊗U(1) covariance:

$$D_{qL\mu} = \partial/\partial x^\mu + \tfrac{1}{2} i g_2 \boldsymbol{\sigma}\cdot\mathbf{W}_\mu + i g_1 B_\mu/6 \qquad (9.15)$$

and a right-handed covariant derivative[37]

$$D_{qR\mu} = \partial/\partial x^\mu \sigma_3 + i g_1 B_\mu |Q| \qquad (9.16)$$

where |Q| is the absolute value of the charge operator (with u eigenvalue 2/3 and d eigenvalue 1/3). The absolute value is used in order to compensate for the minus sign in front of the right-handed tachyon (d quark) term. As a result expression 9.14 becomes

[36] Therefore quark confinement is required in order to have a properly formulated quark sector. Another interaction – the strong interaction – is required for quark confinement. Presently there is only one accepted mechanism for quark confinement – through a non-abelian gauge coupling. (Higher derivative theories with quark confinement are in disfavor.) An additional non-abelian symmetry must be introduced for quarks. As discussed later, SU(3) appears to be a natural choice.

[37] The quark SU(2) coupling constant is, by gauge invariance, required to have the same value as the leptonic SU(2) coupling constant. The U(1) coupling constants are not required to be the same in both sectors and, in fact, are different. The coupling constants here are defined by $e = g_1\cos\theta_W = g_2\sin\theta_W$.

$$\Psi_{qL}{}^{\dagger}(x)\gamma^0 i\gamma^\mu D_{qL\mu}\Psi_{qL}(x) + \Psi_{qR}{}^{\dagger}(x)\gamma^0 i\gamma^\mu D_{qR\mu}\Psi_{qR}(x) \qquad (9.17)$$

Thus the quark sector of the lagrangian[38] (modulo mass/Higgs, and strong interaction, terms) is

$$\mathcal{L}_{\text{quark1}} = \Psi_{qL}{}^{\dagger}\gamma^0 i\gamma^\mu D_{qL\mu}\Psi_{qL} + \Psi_{qR}{}^{\dagger}\gamma^0 i\gamma^\mu D_{qR\mu}\Psi_{qR} \qquad (9.18)$$

$$= \Psi_{qL}{}^{\dagger}\gamma^0 i\gamma^\mu D_{qL\mu}\Psi_{qL} + \overline{\psi}_{uR} i\gamma^\mu D_{qR\mu}\psi_{uR} - \overline{\psi}_{dR} i\gamma^\mu D_{qR\mu}\psi_{dR}$$

$$= \Psi_{qL}{}^{\dagger}\gamma^0 i\gamma^\mu D_{qL\mu}\Psi_{qL} + \overline{\psi}_{uR} i\gamma^\mu(\partial/\partial x^\mu + \tfrac{2}{3} i g_1 B_\mu)\psi_{uR} -$$
$$- \overline{\psi}_{dR} i\gamma^\mu(\partial/\partial x^\mu + \tfrac{1}{3} i g_1 B_\mu)\psi_{dR}$$

where we *provisionally* identify the tachyon as a d-type quark and the Dirac particle as a u-type quark. Our quark sector lagrangian is now the usual Standard Model quark sector lagrangian (modulo mass/Higgs, and strong interaction, terms) except that d quarks are tachyons.

E. The SU(2) gauge field sector has the usual Yang-Mills lagrangian terms, and the B field is a U(1) abelian gauge field.

F. Spontaneous symmetry breaking of gauge symmetry, and of Superluminal covariance, via the Higgs mechanism can be implemented in such a way as to give the quarks, and massive vector bosons, their "known" masses. Since spontaneous symmetry breaking breaks Superluminal covariance of the dynamical equations of motion to Lorentz covariance it is a moot point whether the Higgs sector is manifestly Superluminal covariant or not.

Thus Superluminal covariance of the quark equations of motion generate most of the "unusual" features of the Standard Model.

[38] Note that the gauge fields do not appear with a tachyon equivalent since they are initially massless prior to spontaneous symmetry breaking.

Three Generations and Mass Matrix Issues

The origin of the three generations of quarks and leptons, and the form of the Kobayashi-Maskawa mass matrix[39], are not resolved by Superluminal covariance. Nor is the color symmetry SU(3) of quarks specified. The charges and U(1) coupling constants of the u-type and d-type quarks are also not derived.

A Rationale for an SU(3) Color Group

Although our derivation yielded the general form of the quark sector of the Standard Model it did not appear to determine the nature of the strong interaction binding quarks together. It has long been known that the strong interaction must have a non-abelian symmetry group in order to have quark confinement – a feature required by our derivation and in apparent complete agreement with experimental data.[40] If we require that spin ½ bound quark states exist (assumption 3 earlier in this chapter) as they do, then strict quark confinement would rule out an SU(2) color strong interaction. Therefore SU(3) would be the most minimal non-abelian symmetry group of the strong interactions. Thus there is a rationale for SU(3) color symmetry based on a principle of minimality.

---

[39] M. Kobayashi and K. Maskawa, Prog. Theor. Phys. **49**, 282 (1972).

[40] Evidence is scanty at the present time for five quark states and other exotics that would seem to differ from strict confinement in three quark or quark-antiquark bound states.

10. One Generation Standard Model

10.1 Combined Fermion Sectors

In the preceding chapters we have developed the form of the fermion sectors of a one generation Standard Model based on the Superluminal covariance of the equations of motion. We can thus say that physics beyond the speed of light (the light barrier) plays an important role in the nature of the fundamental physics of sublight phenomena in our universe. It appears we have reached the beginnings of a deeper level of understanding of the nature of the Standard Model.

While some may feel that the regime of superluminal physics is an unwarrented (unphysical) extension of our sublight experience of nature, it is certainly a more conservative theoretical extension than positing additional space-time dimensions, internal symmetries, strings, and superstrings for which no significant evidence yet exists.

It is clear from the discussion in chapter 0 that Black Holes must contain tachyons – faster-than-light particles – and there is much evidence for the existence of Black Holes. Thus we are confronted with the reality of faster-than-light physics and the hypothetical nature of many proposed more fundamental theories for which the Standard Model is a "low energy" approximation.

We conclude this chapter with the derived form of the leptonic and quark sectors of the one generation Standard Model:

$$\mathcal{L}_{SM} = \Psi_L^{\dagger}\gamma^0 i\gamma^\mu D_{L\mu}\Psi_L - \Psi_R^{\dagger}\gamma^0 i\gamma^\mu D_{R\mu}\Psi_R +$$

$$+ \Psi_{qL}^{\dagger}\gamma^0 i\gamma^\mu D_{qL\mu}\Psi_{qL} + \Psi_{qR}^{\dagger}\gamma^0 i\gamma^\mu D_{qR\mu}\Psi_{qR} + \mathcal{L}_{Gauge} + \mathcal{L}_{Higgs} \tag{10.1}$$

where

$$D_{L\mu} = \partial/\partial x^\mu + \tfrac{1}{2} i g_2 \boldsymbol{\sigma}\cdot\mathbf{W}_\mu - i g_1 B_\mu/2 \tag{10.2}$$

$$D_{qL\mu} = \partial/\partial x^\mu + \tfrac{1}{2} i g_2 \boldsymbol{\sigma}\cdot\mathbf{W}_\mu + i g_1 B_\mu/6 \tag{10.3}$$

$$D_{R\mu} = D_{qR\mu} = \partial/\partial x^{\mu}\sigma_3 + ig_1 B_\mu |Q| \qquad (10.4)$$

and where $\mathcal{L}_{Gauge} + \mathcal{L}_{Higgs}$ contain interaction terms with quarks and leptons as well as "free" Higgs and gauge field terms.

10.2 Solution of Parity Violating Form of the Standard Model

The issue of parity violation[41] discovered 50 years ago is now "explained" by the Superluminal covariance of the equations of motion of the unbroken Standard Model lagrangian fermion sector.

[41] T. D. Lee and C. N. Yang, Phys. Rev. **104**, 254 (1956).

11. Higgs Mechanism for Tachyons

The Higgs mechanism is currently the favored mechanism for spontaneous symmetry breaking and to give masses to fermions and bosons. The nature of the free tachyon lagrangian terms,

$$\mathcal{L}_{\text{free}} = \psi_T^{\dagger} i\gamma^0\gamma^5(\gamma^\mu\partial/\partial x^\mu + m_0)\psi_T(x) \tag{11.1}$$

where m_0 is a possible bare mass, requires a Higgs sector that contributes to a tachyon mass through spontaneous symmetry breaking having the general form:

$$\mathcal{L}_{\text{Higgs}} = \tfrac{1}{2}\partial\phi/\partial x^\mu\partial\phi/\partial x_\mu - \psi_T^{\dagger} i\gamma^0\gamma^5\psi_T\phi - V(\phi) \tag{11.2}$$

where

$$V(\phi) = g^2(\phi^2 - (\delta m)^2)^2 \tag{11.3}$$

Note the quadratic term in ϕ – the "mass" term has the negative sign of a tachyon – again showing that tachyons are a feature of modern physics. In the present case the quartic term "stabilizes" the tachyon field which then can be shifted to the minimum of the potential.

The spontaneous symmetry breaking resulting from the potential $V(\phi)$ causes the mass of the tachyon to change to

$$m = m_0 \pm \delta m \tag{11.4}$$

A choice of vacuum state corresponding to the positive sign in eq. 11.4 causes the tachyon mass to increase. A choice of the vacuum state corresponding to the negative sign causes the tachyon mass to decrease, and could cause m to become negative. However this event would not make the tachyon into a normal particle. Rather it would essentially transform the tachyon into its antiparticle.

12. Implications for the "Ultimate" Theory of Elementary Particles

12.1 The Basis of ElectroWeak Theory and the Standard Model

Nature has a way of making very complex phenomena emerge from comparatively simple-looking theories. But how to find the secret of the simplicity! If we look at the vast array of phenomena in Nature, how could we have anticipated that it all follows from the comparatively few terms in the Standard Model lagrangian? Yes, there are some remaining problems – big ones: dark energy, dark matter; and "smaller" ones – unresolved accelerator experimental discrepancies from Standard Model predictions. But at today's accelerator energies the Standard Model does very well.

Perhaps the most important objection to the Standard Model is its hitherto unexplainable form. Why parity violation? Why the form of the ElectroWeak symmetry SU(2)⊗U(1)? Why the form of the symmetry breaking? Why three generations? Why mixing of the three generations? Why the SU(3) strong interaction of the quarks?

In this book we have provided a simple derivation resolving several major questions and making the form of the Standard Model understandable as a consequence of the Left-handed Extended Lorentz group. In particular, we derived

1. The form of parity violation.
2. The SU(2) ⊗U(1) symmetry.
3. The reason for left-handed doublets and right-handed singlets.
4. The need for quark confinement.
5. The form of the Strong interaction SU(3).

We did not derive

1. The charges of particles.
2. The values of coupling constants.

3. The existence of three generations.
4. The mixing of generations.

Thus many features of the Standard Model remain to be understood.

12.2 Is Particle Physics Almost Finished?

Many physicists have discussed the possibility of a "Theory of Everything" or a "Final Theory." Our progress in deriving most major features of the Standard Model raises the hope that an understanding of the remaining features of the Standard Model may be within reach.

However, there appear to be major aspects of the universe, such as dark energy and dark matter, that suggest we have much more to learn. Therefore it is important to continue the program of building larger accelerators and larger "telescopes" to learn more about the universe in the small and in the large. We have learned that they are intimately connected. And we must remember the lesson of the late 19th century when scientists generally thought all major aspects of physics were well understood, and it only remained to clean up a few remaining details. The details, such as black body radiation, the spectra of atoms, and the photoelectric effect, were door openers to the quantum world and to the Standard Model of today.

So, with many "details" still facing us, we can be confident that nature's *secrets are still inexhaustable* and fundamental physics is still just beginning.

This work, which establishes a basis for the Standard Model, and earlier work[42] that proved that Gödel's Theorem implies Nature must be quantum, are the beginnings of a new substratum of fundamental Physics that seeks to provide an understanding of why fundamental Physics theory takes the form that it does.

12.3 Is a "Complete" Theory Possible?

Physicists generally assume that a complete, fundamental theory of physics is possible – a theory that, in principle, describes all physical phenomena. At the level that we are working today, and in the immediate past, it seems that some sort of particle theory describing all particles and interactions could be created. The Standard Model, for instance, is remarkably successful.

However this optimism may well be unfounded. If we look at the sister field of Mathematics, we see major disagreements among mathematicians about the foundations of Mathematics, and, indeed, about what constitutes mathematical proof. There are several theories of the fundamental basis of Mathematics. Since Mathematics is based

---

[42] Blaha (2005b).

on thought (although this is disputed by some mathematicians who feel mathematics grows by investigating new mathematical structures and phenomena "experimentally"), and does not have the open-ended experimental issues of Physics, it seems presumptuous to think that a complete, fundamental theory of Physics[43] can be constructed – especially since such a theory, being mathematical in nature, requires a complete theory of the foundations of Mathematics (which doesn't exist and is not likely to exist for a long time.) A complete theory of Physics is ultimately tied to a complete theory of the Foundations of Mathematics.

12.4 If a Complete Theory of Physics is Developed

If in the distant future a complete theory of Physics is constructed and a complete understanding of the Foundations of Mathematics is developed, then we will be faced with another question of profound importance: Why? Why was this theory the complete theory and not another? Perhaps some principle of self-organization will surface to provide an answer. Perhaps we will have to find a principle of selection, such as Darwin's "Survival of the Fittest" in Evolution theory, and examine all possible physical theories in the light of this principle. Perhaps, the answer is as simple as the Anthropic Principle, "We wouldn't be here if it were otherwise."

More likely, the answer, if there is one, will elude us. One merely has to think of the primitive terms of Euclidean geometry, such as line and angle. With one interpretation of the primitive terms Euclid's geometry emerges. With a different interpretation of the primitive terms non-Euclidean geometry follows.

The understanding of the primitive terms of a complete theory of Physics may well be a major stumbling block. In any case, the study of the primitives will take us out of Physics into the miasma of Metaphysics. And so Physics will ultimately be like Mathematics – a study based on pure thought.

If that is the case, then it is possible that Physics will seriously[44] entertain the question of alternate possible universes based on alternate, consistent complete theories.

---

[43] See Blaha (2005b) for an extended discussion of these issues and also Gödel's Theorems.

[44] There are numerous current efforts to study "many universes" theories. In our view these efforts, while interesting, are premature since we do not known the fundamental theory of our universe or even the form it will take. Thus, for example, a many universes theory based on quantum splitting of universes instant by instant assumes a deeper knowledge of quantum effects, and the nature of reality, than we currently have.

The consideration of "many universes" theories at present is thus a study founded on ignorance and not likely to lead to lasting results – especially in the absence of any means of experimental verification or guidance. Since Newton's times the twists and turns of physical theories have been set by experiments. Theorists are guided by experiment. And many theories that appeared to be natural extrapolations of current theories have been disproven by experiment. A comparison of Physics theory 25 years before 1900 and 25 years after 1900 vividly demonstrate this point.

Such theories would have to be complete (completeness is at present an open question), consistent (mathematical consistency has been shown to be not provable within a mathematical system by Gödel), and sufficiently complex to define an interesting universe.

Thus we see that we are only at the beginning of our understanding of fundamental Physics.

If an ultimate theory of physics is eventually developed that allows us to understand its whys and wherefores, then it will be reasonable to consider alternate physical theories that lead to other types of universes in an intelligent manner. It might even be possible to create "bubble" micro-universes within our universe to experimentally verify these alternate theories. The prototypical analogy for bubble universes is a Black Hole. A bubble universe would have a horizon that allows its interior to have a different physics, and a rapid "time" evolution. And an experimenter would have probes to measure the dynamics within the bubble universe to confirm/refute an alternate physics.

This brief consideration of bubble universes raises an interesting question: Is our universe a bubble universe in some grand experiment? At the moment the answer, if there is one, is metaphysical in nature.

REFERENCES

Bjorken, J. D., Drell, S. D., 1965, *Relativistic Quantum Fields* (McGraw-Hill, New York, 1965).

Blaha, S., 2003, *A Finite Unified Quantum Field Theory of the Elementary Particle Standard Model and Quantum Gravity Based on New Quantum Dimensions™ and a New Paradigm in the Calculus of Variations* (Pingree-Hill Publishing, Auburn, NH, 2003).

Blaha, S., 2004, *Quantum Big Bang Cosmology: Complex Space-time General Relativity, Quantum Coordinates, Dodecahedral Universe, Inflation, and New Spin 0, ½, 1 & 2 Tachyons & Imagyons* (Pingree-Hill Publishing, Auburn, NH, 2004).

Blaha, S., 2005a, Quantum Theory of the Third Kind: A New Type of Divergence-free Quantum Field Theory Supporting a Unified Standard Model of Elementary Particles and Quantum Gravity based on a New Method in the Calculus of Variations (Pingree-Hill Publishing, Auburn, NH, 2005).

Blaha, S., 2005b, The Metatheory of Physics Theories, and the Theory of Everything as a Quantum Computer Language (Pingree-Hill Publishing, Auburn, NH, 2005).

Blaha, S., 2005c, *The Equivalence of Elementary Particle Theories and Computer Languages: Quantum Computers, Turing Machines, Standard Model, Superstring Theory, and a Proof that Gödel's Theorem Implies Nature Must Be Quantum* (Pingree-Hill Publishing, Auburn, NH, 2005).

Blaha, S., 2006, *A Derivation of ElectroWeak Theory based on an Extension of Special Relativity; Black Hole Tachyons; & Tachyons of Any Spin.* (Pingree-Hill Publishing, Auburn, NH, 2006).

Bogoliubov, N. N., & Shirkov, D. V., Volkoff, G. M. (tr), 1959, *Introduction to the Theory of Quantized Fields* (Wiley-Interscience, New York, 1959).

Cottingham, W. N. and Greenwood, D. A., 1998, *An Introduction to the Standard Model of Particle Physics* (Cambridge University press, Cambridge, UK, 1998).

Heitler, W., 1954, *The Quantum Theory of Radiation* (Oxford University Press, London, 1954).

Huang, K., 1992, *Quarks, Leptons & Gauge Fields Second Edition* (World Scientific, River Edge, NJ, 1992).

Huang, K., 1998, *Quantum Field Theory* (John Wiley, New York, 1998).

Kaku, M., 1993, *Quantum Field Theory* (Oxford University Press, New York, 1993).

Misner, C. W., Thorne, K. S., Wheeler, J. A., 1973, *Gravitation* (W. H. Freeman, New York, 1973).

Sakurai, J. J., 1964, *Invariance Principles and Elementary Particles*, (Princeton University Press, Princeton, NJ, 1964).

Streater, R. F. and Wightman, 2000, A. S., *PCT, Spin and Statistics, and All That* (Princeton University Press, Princeton, NJ, 2000).

Turnbull, H. W. and Aitken, A. C., 1961, *An Introduction to the Theory of Canonical Transformations* (Dover publications, New York, 1961).

Weinberg, S., 1995, *The Quantum Theory of Fields Volume I* (Cambridge University Press, New York, 1995).

Weinberg, S., 1996, *The Quantum Theory of Fields Volume II* (Cambridge University Press, New York, 1996).

INDEX

About the Author

Stephen Blaha is an internationally known physicist with extensive interests in Science, the Arts, and Technology. He received his Ph.D. in Theoretical Physics from Rockefeller University (NY). He has written a highly regarded book on physics, consciousness and philosophy – *Cosmos and Consciousness*, a book on Science and Religion entitled *The Reluctant Prophets*, a book applying physics concepts to the history of civilizations, and books on Java and C++ programming. He developed a mathematical theory of civilizations that is described in *The Life Cycle of Civilizations*. Recently he completed a major new study of Cosmology: *Quantum Big Bang Cosmology: Complex Space-time General Relativity, Quantum Coordinates, Dodecahedral Universe, Inflation, and New Spin 0, ½, 1 & 2 Tachyons & Imagyons*. He has served on the faculties of several major universities. He was an Associate of the Harvard Physics Faculty for twenty years (1983-2003). He was also a Member of the Technical Staff at Bell Laboratories, a member of management at the Boston Globe Newspaper, a Director at Wang Laboratories, and President of Blaha Software Inc and Janus Associates Inc. (NH). Dr. Blaha is noted for contributions to elementary particle theory, mathematics, condensed matter physics and computer science.

Among other achievements he was a co-discoverer of the "r potential" for heavy quark binding developing the first (and still the only demonstrable) non-abelian gauge theory with an "r" potential; first suggested the existence of topological structures in superfluid He-3; first proposed Yang-Mills theories would appear in condensed matter phenomena with non-scalar order parameters; first developed a grammar-based formalism for quantum computers and applied it to elementary particle theories; first developed a new form of quantum field theory without divergences (thus solving a major 60 year old problem that enabled a unified theory of the Standard Model and Quantum Gravity without divergences to be developed); first developed a formulation of complex General Relativity based on analytic continuation from real space-time; first developed a generalized non-homogeneous Robertson-Walker metric that enabled a quantum theory of the Big Bang to be developed without singularities at $t = 0$; first generalized Cauchy's theorem and Gauss' theorem to complex curved multi-dimensional spaces; first developed a physically acceptable theory of faster-than-light particles – tachyons – of any spin; first showed a universe with three complex spatial dimensions has an icosahedral symmetry; first developed the form of the composition of extrema in the Calculus of Variations; first suggested that inflationary periods in the history of the universe were not needed; first proved Gödel's Theorem implies Nature must be quantum, and first developed a quantitative harmonic oscillator-like model of the life cycle, and interactions, of civilizations.

Blaha was also a pioneer in the development of UNIX for financial and scientific applications, database benchmarking, and networking (1982); in the development of Desktop Publishing (1980's); and developed a hybrid shell programming technique (1982) that was a precursor to the PERL programming language. He received Honorable Mention in the Gravity Research Foundation Essay Competition in 1978, and was nominated for three "Awards for Technical Excellence" in 1987 by PC Magazine for PC software products that he designed and developed. His email address is sblaha000@yahoo.com.

www.ingramcontent.com/pod-product-compliance
Lightning Source LLC
Chambersburg PA
CBHW081546220326
41598CB00036B/6577
* 9 7 8 0 9 7 4 6 9 5 8 7 7 *